ANN CLIFF has been a lifelong gardener and farmer, starting many years ago with her family's smallholding ventures in England and Wales. After agricultural college she worked as a UK farm training organiser and wrote about farming in magazines and for government and commercial publishers. She hosted a Yorkshire Television series on self-sufficiency and ran regular weekend training courses that included beekeeping.

Ann has written books and articles on rural themes for more than thirty years, both in the UK and in Australia. (Her historical novels are set in the nineteenth-century countryside.) She writes for several magazines, including *Earth Garden* and *Grass Roots*, and has contributed pieces on beekeeping for *Town and Country Farmer*.

For the past twenty years, Ann and her husband Neville have lived on an 80-hectare farm in Gippsland, Victoria, on the edge of a state forest. They produce Angus beef for the local market and grow as much of their own food and fuel as possible. The whole farm, including bees and the food garden, is managed on organic principles with the aim of optimum rather than maximum production.

Over the years, Ann has been an 'intermittent beekeeper' and, after a few bee-less years, is currently very happy to hear the sound of bees again in the garden. She and Neville are concerned about the survival of bees in today's world and they hope for a more bee-friendly future.

THE
BEE BOOK

Beekeeping Basics | *Harvesting Honey*
Beeswax, Candles and other Bee Business

ANN CLIFF

MANNA PRESS
Melbourne

This edition first published in 2010 by
Manna Press (an imprint of Manna Trading
Pty Ltd, formerly Aird Books)
www.hylandhouse.com.au

National Library of Australia Cataloguing-in-
Publication entry
Author: Cliff, Ann.
Title: The bee book / Ann Cliff.
Edition: 1st ed.
ISBN: 9780947214609 (pbk.)
Notes: Includes index.
Subjects: Bee culture--Australia.
Dewey Number: 638.10994

Edited by Bet Moore
Design and layout by Pauline Deakin,
Captured Concepts
Printed by Bookbuilders, China

CONTENTS

ACKNOWLEDGMENTS

I am very grateful for the help of many people and thousands of bees in the preparation of this book. Beekeepers are friendly folk and they have tried to keep me on the right flight path, but any mistakes are mine.

Bill Ringin, secretary of the Gippsland Apiarists Association, some of whose colonies are currently having to cope with the variable climate on our upland farm, has been with me from the start of this project. I owe Bill and his wife Jeannette a great deal for their support and practical information. Bill's knowledge of bees is extensive and he enjoys sharing it with others. Bill also supplied the photograph in chapter 8, of a brood killed by AFB.

Bob Owen of Bob's Beekeeping Supplies in Eltham, Victoria, read the draft and delved into some byways of science to add interest and depth to the account. I am grateful to Bob for his insight and experience and his gift of precious time.

Thanks also go to Nic Moore, a hobbyist beekeeper in the Otway Ranges, Victoria, who allowed us to take many photographs (acknowledged where they have been used), and whose comments on the manuscript were gratefully received.

Rod and Karen Palmer were generous with photos of their native bees, as were Russell and Janine Zabel. Dave Wilson of the Amateur Beekeeper's Association of NSW went to great lengths to help with photos, as did several other ABA members.

Very many thanks go to the busy specialist who read the chapter on bee diseases and made several helpful suggestions.

John and Marian Cardell gave me generous access to their library.

For my initiation into the bee world I will always be grateful to Bill Bielby, a bee adviser who is still remembered by Yorkshire beekeepers. He was the custodian of bees at Fountains Abbey before he went to live in New Zealand.

The main photograph on the front cover, of a bee collecting borage nectar, is from Jen Owens. The same photograph is used on page 46 together with a couple from Allen Gilbert, whilst Penny Woodward provided the bees on lavender and thyme on page 47 and on borage on page 102, as well as the main garden pictures on pp 54-5 and 56-7.

CHAPTER 1

Introduction

A short history

One of the most ancient of human food sources is honey. Hunter-gatherer populations all over the world have raided the nests of wild bees for thousands of years. There is a cave painting in Valencia, Spain, depicting a figure robbing a hive for honey, surrounded by very large bees. The usual outcome of robbing wild nests is the destruction of the bees' home and also the eggs and larvae of the next generation.

Eventually, people started to make artificial beehives and so to 'domesticate' the bees. Early hives were often hollow logs, or made of woven straw, possibly with clay added, wood or pottery. Ancient Egyptian records make many references to bees and sealed pots of honey have been found in the tombs of the Pyramids. For centuries it was still necessary to destroy the hive in order to take the honey.

Monasteries in the Middle Ages were probably the most important beekeeping sites; monks harvested honey for their diet and wax for their church candles, and apparently they liked a drop of mead as well. In the colder countries of Europe where grape growing is difficult, there is a long tradition of fermenting honey to make alcoholic drinks.

Naturally enough a great deal of folk lore grew up with the culture of bees, since their lives are so mysterious to an external observer. In the eighteenth century bees began to be studied scientifically. In France, Reaumur made an observation hive with glass walls and watched the queen bee laying eggs, but no one knew how they were fertilised until much later when a Swiss scientist, Francois Huber, discovered that queens are fertilised by drones.

Huber (1750–1831) was blind from the age of 20, but he employed a secretary to be his eyes. He made some important discoveries. He built a glass-walled hive with sections that could be opened to look at individual wax combs. Over many years he observed and recorded, dissecting bees under the microscope and establishing the basic facts of bee biology. It is amazing to think that, even without the use of his sight, he achieved so much.

By the end of the eighteenth century, the wastefulness of killing colonies of bees for their honey had prompted the development of a better type of hive, with wooden frames on which the bees could build their combs. The combs could be taken away without damaging the colony. The moveable comb hive evolved in the nineteenth century.

The famous beekeeper whose name is associated with the development of the modern hive is the Reverend Lorenzo Langstroth (1810–95) in the USA. Based on Huber's work he developed a series of frames inside a rectangular wooden box, designed in such a way that the bees would build honeycombs on the frames without glueing them together, so they could be removed individually. Langstroth's discoveries were published in his book, *The Hive and the Honey-bee* (1853). The Langstroth hive is still standard in many parts of the world and all modern hives are made on the same principle, with small regional variations. The moveable comb hive made beekeeping much easier, resulting in the rise of commercial beekeeping.

Honey bees were first brought to Australia from England in 1810 by a parson, Samuel Marsden, but they did not survive. The second importation of bees was in 1822 and this was successful. Since then Italian, Carniolan and other bees have been brought in and most of our managed colonies are of Italian bees. In New Zealand, beekeeping started in the 1850s, soon after European settlement.

Keeping Bees

Beekeeping is a wonderful hobby for people who like to have access to a part of the natural world and produce some of their own food.

Bees make a good addition to a farm, but broad acres of your own are not necessary; the free-ranging bees will forage for themselves. Suburban and even urban beekeeping is quite feasible in many places. Frustrated 'city farmers' are taking advantage of the 'greening' of townscapes and the many parks and gardens. In fact, it has been suggested that bees quite like city life (where the temperature is higher than in the surrounding countryside) if they can live well above traffic pollution. There is a hotel in Toronto with three hives on its thirteenth-floor rooftop garden, where they grow herbs and vegetables for the hotel's restaurants.

Bob Owen, who sells beekeeping supplies, says that some of his customers live in inner-city Melbourne and keep their hives on the garage roof. He tells them that cities are a good place to keep bees since there are plenty of European flowers around that the bees have been genetically selected to use. Although, given the choice, honey bees may prefer European flowers, see chapter 5, Food for Bees, for an overview of indigenous species like Spotted Gum, Banksia, Tea Tree and Leatherwood, for example, as excellent sources of pollen and nectar.

Once your bees are established they don't place excessive demands on your time, but you will need to keep an eye on them. Unlike most livestock, bees don't need daily attention, but there is essential and quite strenuous work to be done at certain times of the year. Spring and summer are relatively busy, although in winter the bees are best left to themselves, with adequate supplies of food and water. There is a great deal to learn about these interesting, hard-working insects.

To start, join a beekeepers' association and find out what is involved in the ancient craft, art and science of beekeeping. You will soon be keenly interested in these fascinating creatures and their complicated lives. Some people get hooked unintentionally. During the war, when sugar was scarce, Dr Eva Crane was given a hive of bees as a wedding present. It turned out to be a life-changing gift: over the next 50 years or so she became a world authority on bees.

Even if you decide that beekeeping is not for you, learning about the processes that produce honey and beeswax is fascinating. Once you have a little knowledge about bees, you'll never take honey, or the seasonal pollination of your pumpkins or fruit trees, for granted.

'What if the bees don't like me?' Initially, some people are understandably nervous, scared to approach a hive containing 60,000 little insects that sting. But there are ways to minimise stings and their effect, and ways in which to dress (and smell) so you don't provoke the bees.

Of course, knowledge is vital. You could attend a course – various TAFE and agricultural colleges run courses on beekeeping (see Further Information). But you can't beat practical experience, and the more you find out first-hand, the better. Most apiarists (the proper name for beekeepers, derived from *Apis*, Latin for 'bee') are happy to share their knowledge and enthusiasm.

The health-food store in your neighbourhood may stock local honey and you could give the producer a call. Bob Owen says, 'I have suggested this to many of my customers in more remote places and they have all been successful in finding a mentor.'

Beekeeping plays a vital role in our countryside, but beekeepers, like bees, are in decline. Could you help the environment by becoming a beekeeper?

The arguments in favour of a hive of bees include their products, honey and wax, and also their effect on the environment. Up to 60 % of crops and many wild plants depend on bees for pollination and therefore production. The bad news, which is causing much concern, is that there is a worldwide decline in bee populations. Climate change, agricultural chemicals, stress and disease are thought responsible, but nobody seems to know for sure.

At the same time, there is more pressure on bees in commercial hives to perform. Some monoculture crops like almonds demand a large number of bees at the right time for pollination and are thus dependent on commercial beekeepers. There is an increasing demand for pollinators for biofuel crops like palm oil and canola. Wild native bees and other insect pollinators can do some of the work, but modern crops demand large numbers of pollinators during the relatively short period when pollination is effective. (In 2009, 52,000 hives were needed from four states for the almond crop belonging to a large organisation in receivership. Had the bees not been taken there when the trees were in flower, at a cost estimated at more than three million dollars, the crop would have failed. Fortunately the financial experts saw the light and let the bees in.)

When bees pollinate almonds and other crops, they often lose condition. Numbers in the hives decline and the bees lose strength. A monoculture is not natural to them and it may provide high-moisture nectar, which means more work, or low-protein pollen, which gives them inferior food. Beekeepers therefore often feed their bees when they are pollinating, and nurse them back to strength afterwards by giving them greater variety.

Threats to bees' welfare in the modern world

Agricultural insecticides are believed to have had a dramatic, negative effect on bee populations. The bees are killed along with the targeted species.

Extreme weather conditions are predicted to become worse and more frequent with global warming. The very wet summer of 2008 in the UK, for example, had a disastrous effect on bee populations there, and drought affects the flowering time of plants and may cause eucalypts in Australia to delay flowering until the rain comes. Climate worldwide has become less predictable. Bees are not able to forage in severe weather and they use their stores of honey; without food, they soon die. Bushfires and floods are damaging to people, flora and fauna, and also affect bees.

Some pests and diseases of bees have apparently spread through international trade and include exotic pests like the Varroa mite, which we hope won't come to Australia, although it has already been detected in New Zealand and PNG. At the moment Australian bees are relatively healthy and are in demand in the UK and other European countries experiencing a shortage of bees. But with so much movement of people and goods around the world, disease thumbs its nose at frontiers and no country can expect to be spared from infection. The fear of 'pandemics' of one kind or another often dominates the headlines.

Colony collapse disorder reported in 2009 by the ABC is a mysterious, new phenomenon in the USA and the UK. The bees do not show any sign of disease, they just disappear. The disorder may well result from a combination of causes – the stresses of changing weather and current agribusiness practices (large areas of monoculture, large numbers of bees transported long distances). The bees may lose their resistance to disease and, possibly, their will to live. There is also speculation that the bees' homing instinct, the ability to locate their hive, is affected, causing disorientation and hence loss to the colony. Varroa and Nosema, certain nicotine-based pesticides, and the Israeli Acute Paralysis Virus have all been suggested as possible causes.

In parts of New Zealand, Varroa has killed most native and feral bees, leaving only managed hives to collect nectar and pollen. Because there is now little competition from feral bees for nectar, the amount of honey produced by managed bees has significantly increased.

But all these worries apart, beekeeping remains an important industry with a significant contribution by hobbyists. There are estimated to be

Bee All & End All

The bee holocaust myth is just another example of our strange yearning for catastrophe

THE WORLD IS GOING TO END IN 2012, apparently, hopefully, just before the start of the Olympic Games (or ol...

Outbreak of deadly bee mite 'inevitable'

into the population," said Dr Denis Anderson, a CSIRO virologist ... specialist on

limited to the wellbeing colonies. In Australia, a ing industry is built c pollination and transp ... around the

'Heater' honey bees have hives humming in harmony

Scientists have lifted the lid on the social make-up of colonies.

THE secret of honey bees' success has been found living deep inside their hives — a special type of bee that acts as a living radiator, warming the nest and controlling the colony's complex social structure.

The "heater bees" have been found to play a crucial, previously unappreciated, role in the survival of honey colonies.

Using new technology that allows scientists to "see" the temperature inside the hives, researchers have been able to ...

They have found that these specialised bees, whose body temperatures are considerably higher than other bees in the colony, not only keep the hive warm but to control the social make-up of a colony.

Bees, and other social insects such as ants, share jobs in a colony, with each individual having a specific role ...

... of labour ... they ...

for maintaining the temperature of the brood nest in a hive, where young bees, known as pupae, are sealed into wax cells while they develop into mature bees.

The scientists discovered that the heater bees work to subtly change the temperature of each developing pupae by about one degree and this small change determines what kind of bee it will become.

Those kept at 35 degrees turn into the intelligent forager bees that leave the nest in search of nectar and pollen. Those kept at 34 degrees emerge as "house ...

Wurzburg University in Germany, said: "The bees are controlling the environment they live in to make sure they can fill a need within the colony.

"Each bee in a colony performs a different profession — there are guard bees, nest-building bees, brood-caretaking bees, queen-care ...

ops and the likelihood of the role it will fulfil when it emerges as an adult."

Thermal imaging cameras show how heater bees warm the nest. By beating the muscles that would normally power their wings, heater bees increase the temperature of their bodies to 44 degrees, nearly 10 degrees hotter than a normal bee.

They then ...

size of the hive, also press themselves against individual cells to top up the temperature of each pupae.

Professor Tautz said: "The old idea was that the pupae in the brood nest were producing the heat and bees moved in there to keep warm, but what we have seen is there are adult bees who are responsible for maintaining the temperature.

Temperature is know ...

To bee or not to bee, this is what the buzz is all about

pollinate the schemes result in "virtually no crop being set this year.

KordaMentha mainta seven almond and seve projects — the almond ...

... for bees — about ...

Always carrying a touch of mystique, bees have become a bit controversial in recent decades.

600,000 hives in Australia and more than 300,000 hives New Zealand (in 2007). They produce an average of 67 kg of honey per hive, with commercial operators producing 200 kg or more.

Considerations for intending beekeepers

You really need hands-on experience (it's worth repeating) with an established beekeeper before investing money and time in bees. Most are enthusiasts and happy to pass on their knowledge to someone who is willing to observe quietly, dress appropriately and not upset the bees. Some of the training courses have a practical component.

Your location is important if you intend to keep the hive or hives at home. If your home area is not bee friendly, you will need to find a place for them where they will be secure and able to go about their business.

Suburban gardens are a rich source of nectar and many Australian trees yield good honey. But bees and people do not mix and neither do bees and horses, so there is a need to keep people away from the hives. There can be complaints if the bees' flight path coincides with human walkways. To avoid this, place an obstacle, fence or hedge, in front of the hive to force the bees to fly upwards before they depart to forage. Worker bees can travel more than 2 km during foraging flights, but will normally seek nectar and pollen as close to the hive as possible. Once out of the hive they spread out and gain altitude, so you don't need to be that far from people.

The bees also need to be safe from vandals, dogs and other animals that might upset the hive.

There are hazards even in farming areas, where bees might be expected to be safer. Crop spraying can be lethal, farm animals may tip over the hives or trees may blow down on them. Orchardists will usually welcome hives, but if you help out a grower with pollination you will need to find another source of nectar once the job is done, that is when all the trees have shed their blossom and set fruit.

Bee stings are inevitable if you keep bees, although they can be minimised by protective clothing and good handling techniques. If you are thinking of becoming a beekeeper, it may be as well to find out before you start whether in fact you are sensitive to bee venom. Some beekeepers suggest that you visit hives and make sure you get stung – on the arm for preference. Stings on the face are obviously more dangerous, especially near the mouth or eyes, which is why veils are worn even by people who approach hives with bare hands and arms.

A few people are severely allergic to bee stings and for them there is a danger of anaphylaxis. If you are one, beekeeping is obviously not for you. For most people the first sting or two may be painful, but they develop antibodies and are less affected over time. I have heard of a beekeeper for whom the opposite applied. His reaction to bee stings became more severe and in the end he had to give up keeping bees.

Speed matters. If you are stung, try to get rid of the sting and the bee as fast as you can with a scraping movement of your fingernail or the hive tool and, if you are quick, the bee will not have had enough time to pump much poison into you. The bee leaves a barb embedded in your skin, still pumping poison, and if you rub the spot, more venom will be injected. A cold compress may help to ease the pain and so may anti-sting cream.

The beekeeper aims to reduce the number of stings by trying to minimise the triggers that cause bees to attack: working in the right conditions in the middle of the day, choosing a queen with a docile temperament, having plenty of smoke in the smoker, moving quietly and smoothly and avoiding bumping the hive or squashing bees. Clumsy behaviour is, as a rule, not tolerated by these little insects.

Scientists have shown that the alarm pheromone produced by bees when they are upset is the same chemical produced by ripe bananas, isopentyl acetate. They release the strong scent as they sting, which provokes the other bees to sting too. So the rule is not to eat a banana when you are working around the hive.

Another consideration is – how much honey will I harvest? The costs of setting up can be worked out, but the yield of honey is not easy to predict. Seasons vary and with them the flowers, quite apart from the effect of weather on the bees themselves.

They don't like extremes of weather. The decline in commercial bee-keeping may be partly due to the fact that honey production is even more of a gamble than other forms of primary production (and that is saying something!). As an example of the variation in experience, in autumn 2009 Canberra apiarists were recording a 'bumper crop' of honey, while in the large areas of Victoria affected by drought and bush fires there was obviously a massive decline in bee fodder and beekeepers were asking for further access to placing hives in state forests. (There are currently about 3000 designated bee sites on public land in Victoria.) Through the following winter, local honey was scarce in these areas.

Of course, as a hobby beekeeper you are not setting out to make a living from your bees. Professional apiarists have a busy time, often moving their hives on special trailers to predicted honey flows. (You won't see them often because like smugglers, they move at dead of night.) Amateur beekeepers will be able to take the bad years with the good and will not have invested too much money or effort in the enterprise, although it goes without saying that once you commit to a hive of bees you have a duty of care to look after them and to safeguard other people from them too.

As older beekeepers retire there are fewer young people learning the skills, and Australian apiarists have registered their concern about the lack of new people entering the industry. The last few years have seen a revival of interest, however, and there are now beekeeping courses in most states. The Queensland DPI records that beekeeping is becoming increasingly popular in towns and cities and over 2000 households have registered hives. So perhaps the amateurs, the hobby beekeepers, will have a useful future role.

It seems to me that all the problems faced by bees and their keepers point to the need for an environmentally friendly beekeeping system. The insects have been 'domesticated' for thousands of years, but seem to be more vulnerable now then ever before. This is why many experienced apiarists are now thinking that older, more traditional methods may be best. That is, they will intervene only when necessary and otherwise leave the bees to get on with their work after giving them the optimum conditions, as far as possible. It seems obvious that bee colonies are 'free-range' and free spirits and have to be left to their own devices for much of the time. Bees will also benefit from more sustainable farming practices with fewer poisons, as organic principles become more widely applied.

CHAPTER 2

Bee Species and Behaviour

How doth the little busy bee
Improve each shining hour
And gather honey every day
From every opening flower.

Isaac Watts

First of all, meet the bees. Before we start, it helps to know a little about the insects, their biology and their life cycle. The more we know about them, the easier it is to work with them and to work with nature, the aim of all sustainable enterprises.

Species

All commercial honey bees belong to a species of advanced social bee, *Apis mellifera*, the honey bee. The two other groups are solitary bees and social bees (e.g. bumblebee). They have lived in warm temperate and subtropical forests for 30 million years. According to recent research, they originated in Africa and have come out of there at least three times, to eastern and western Europe and to South America (see 'Thrice Out of Africa', *Science Magazine* 2006).

Three subspecies of honey bee are generally available in Australia; they all originate from Europe and they can interbreed. They are:

Apis mellifera ligustica, the Italian bee. This is the most common managed honey bee in Australia. The workers have between two and five yellow bands on the abdomen. They start working quite early in spring and their temperament varies. They develop large colonies.

Apis mellifera carnica, the Carniolan bee. The bees have brown bands of short hair on the abdomen and the workers look grey. They are the most docile of the three, but have a tendency to swarm more often. Carniolan bees overwinter in a small colony and then breed fast in spring to get up to summer strength.

Apis mellifera caucasia, the Caucasian bee. Darker than the others, it has a longer tongue so it can reach into flowers that the others can't. Caucasian bees are docile and prefer cooler conditions. They like to fill up gaps with a great deal of bee glue, propolis, which can be a drawback to the beekeeper.

The strain of bee within the species is the most important consideration. You can change the type of bee you keep when selecting a new queen, as all the workers will be descended from her. Often, beekeepers faced with aggressive bees will introduce a docile queen with the intention of calming things down.

It is a sad fact that when bees are desperate and conditions are poor they may turn to crime. They become marauding and aggressive and can give you and your neighbours trouble, stealing honey, attacking people and

A Caucasian queen bee with Italian bee workers. The queen is bottom left, bigger and darker than the workers, and has a splash of white on the thorax for identification. Her introduction means that the next generation of workers will be Caucasian and not Italian.

possibly spreading disease. Robber bees are the reason why the bees themselves have guards on the door to keep the criminal element out, and the reason why beekeepers must take precautions never to leave honey or sticky frames out in the open, nor allow bees access to storage areas for honey and equipment. It is illegal in all states to leave honey or equipment accessible to bees. The Code of Practice for beekeepers in each state warns you to beware of robber bees. (Have a look at the Code of Practice on your state department's website – see Further Information.)

Apart from the fact that you may lose your honey, robbing is a major way in which disease can spread. Once disease gets into a colony and weakens it, that colony may not be able to defend its own hive from robber bees.

The big, heavy, hairy bumblebee is also on the black list in most states. Bumblebees have established themselves in Tasmania and in New Zealand, but are not supposed to be on mainland Australia. If you think you see one that has sneaked in, you should report it. They are striped and hairy, so big they can't hide – up to 35 mm long – and they sound like a small tractor. Their crime is to compete with native fauna for nectar and they may also encourage exotic weeds that currently have no efficient pollinator – the so-called 'sleeping weeds'.

It is interesting to read (June 2009) that the short-haired bumblebee was introduced to New Zealand towards the end of the nineteenth century, to help with the pollination of clover crops. It has since died out in Britain and is to be re-introduced there from New Zealand.

Native bee species

Australian native bees now have a support group, Aussie Bee (www. aussiebee.com.au) and there is a native bee research centre in the Blue Mountains. Over 1500 species have been recorded, of many different sizes and colours. Many are solitary bees living in burrows in the ground. There are about 10 species of native social bee that do not sting and they have caused much interest. They are small. Trigona is about 4 mm long, whereas the European honey bee is about 12 mm long. Native honey is a wonderful bush food, but tiny bees live in small colonies and so the yield of honey, called sugarbag, is small. Aboriginal communities around the world collect this type of honey from wild bees.

It is now over 20 years since the 'domestication' of some species of native bee was first tried in Australia. Although small, like the bees themselves, there is now an established cottage industry in the warmer states where the suitable species of bee live.

Of the 1500 different species of bee native to Australia, 10 species are social bees living in a colony, and will therefore lend themselves to beekeeping. Of those, the three main species are:

- *Austroplebeia australis*, a timid little bee that lives in southern and western Queensland.

- *Trigona hockinsi* is from northern Queensland and needs the heat.

- *Trigona carbonaria* lives on the coast, from Queensland down to Bega in NSW. This is probably the most versatile species.

These little bees do not sting. Stingless beekeeping sounds wonderful; they start with a great advantage as far as humans are concerned. (Even so, they bite if provoked and can get into your ears and nose.) It's quite feasible to keep them in the warmer states of Australia; these bees do not like the cold. They also have a different, smaller hive configuration that makes honey removal more difficult.

Stingless beekeeping is not recommended in Tasmania, Victoria, ACT or South Australia, as the southern states are not in the natural geographic range of these species. In Western Australia these bees occur only north of

Native bees (Russell and Janine Zabel)

the Hamersley Ranges, and they are not available commercially. In this state, the authorities are against the introduction of bees south of these Ranges, in case they upset the balance of plants and pollinators.

'Bob the Bee Man' in Highvale, Queensland, advertises that he is trying to save as many colonies of native bees as he possibly can. They are threatened by earth moving, timber felling and some types of farming. He says that little is known about native bees and we do not yet know their full potential. His web page is addressed to 'anyone who may come into contact with the stingless native bees of Australia'. He asks people to contact him when they encounter threatened sites of native bees (for example, if you find a colony when felling a tree). He gives advice on first aid at such sites and, if possible, he will relocate the colony, or supply a native-bee rescue kit. He is boxing some colonies to work out the optimum hive design.

Native bees are of increasing importance in Queensland and New South Wales, kept for a hobby by interested people. Some people want them as pollinators and some to look pretty on the garden flowers, while others like the idea of native honey as a real bush food. The insects would make a wonderful teaching resource in a school garden.

Using native bees as a resource may be a way of conserving at least some of the species for the future; they obviously have a role in pollinating native plants and can play a part in regeneration projects. Conservation of the bees themselves is now important, because they are declining in numbers as the dead and hollow trees they live in are cleared. Many landowners do not leave trees to age and die. They may also be threatened by chemicals in the environment, as are all bees.

Box hives are now common, but log hives, because of their greater insulation, are better for the bees in cooler areas such as the south coast of NSW.

Native bee enthusiasts and experts all agree that wild honey should not be harvested, as it can damage the nest and harm the colony.

Because they are mainly tropical insects and don't have to face long cold winters, native bees need to produce only small amounts of honey, which they store in resin pots, little spheres about 12 mm across. The honey, called sugarbag, is more strongly scented and flavoured than that of the European honey bee. It is the more appreciated because a native beehive will produce only about one kilogram of honey a year. Keeping these bees will be fascinating, but it's not a commercial proposition.

You can now buy stingless bees that can be kept in small hives (see Further Information). They are used to pollinate macadamia nut trees and tropical fruit trees, but their range is only about 500 m, so they need to be close to the crop.

There are reports of mixed success with native bees. They are sensitive to cold and also to extreme heat, as are all bees. They don't fly at such low temperatures as European bees; they only work in a temperature range of 18–35°C. They can be protected from extreme temperatures by covering the hive with a polystyrene box on very hot or very cold days.

Usually the species can exist side by side, but if there is competition for nectar sources, the bigger bees will win.

A native beehive has been developed by researcher Tim Heard in Brisbane, who has been responsible for much of the interest in these little creatures (www.sugarbag.net). He made a box like a miniature beehive, in two halves so that strong colonies could be split in two.

Russell and Janine Zabel describe themselves as keepers of Australian stingless native bees and they maintain an informative and friendly website. The bee boxes they sell are made to be mounted on a star picket post. They say the post should be kept oiled to deter ants and spiders. They dispatch hives and logs of bees to customers from mid-October.

Native bees have the same life cycle as other bees and their management is not all that different, except that they seem to be more fragile. They need clean water close to the hive and a year-round supply of pollen and nectar.

Some flies are enemies of native bees. One hoverfly lays eggs on the outside of the box and the maggots prey on the hive. Another enemy is a grey fly, small enough to get in through the hive's entrance.

Another difference with Trigona native bees is the curious habit of fighting swarms. Thousands of worker bees congregate outside a hive and fight each other. Meticulous research (Australian Native Bee Research Centre) has established by DNA sampling that the fighting swarms they investigated were all caused by bees from one hive attacking another. The defending hive was never more than two metres from the swarm. Unfortunately, sometimes in the fray the bees will kill members of their own hive. This event has distressed some beekeepers, but the researchers say that it's usually not a calamity and the hive may afterwards be stronger, as the stronger bees will win.

A hive for native bees (Russell and Janine Zabel)

According to the Zabels' fact sheet, native bees will harvest the seeds of a eucalypt called Cadagi (*E. torrelliana*), a large tree that lives on the edge of tropical rainforest. The seeds open in January and February and the bees take the seeds back to the hive because they use the resin on them as material to build nests and to seal cracks. The danger is that sometimes the entrance to the hive can get blocked up with seeds and the colony can die, so beekeepers need to clear the hive entrance if this happens.

If you are lucky enough to live in one of the tropical paradises where the little native bees can flourish, you might like to consider keeping them.

Feral bees

Feral bees are controversial. These enterprising creatures are bees that came from domesticated stock, from managed hives, and escaped to lead their own lives. They have swarmed from another colony, feral or managed, and the swarm has escaped capture or destruction.

Feral bees are seen as an environmental threat in many areas. They are not native and they could harbour disease that might affect managed hives and native bees. In New South Wales they have the label 'Key Threatening Process'.

In some areas of New Zealand feral bees have been eradicated through selective use of poisoned sugar baits and in Western Australia the same thing has happened. It is feared that they upset pollination in native plants and compete with native fauna such as native bees, honeyeaters and black cockatoos. (Some scientific papers put feral bees and domesticated bees in the same category.) Feral bees are often aggressive, which does not help their image. And, of course, they can easily spread pests and diseases.

A colony of feral bees, happily living and working in a public park tree in Geelong, Victoria.

On the other hand, feral bees can be important for crop pollination, and they have some friends. A New Zealand scientist (University of Otago's director of genetics, Peter Dearden) has been reported as saying that feral honey bees are responsible for most of the pastoral, horticultural and crop pollination. He said that in the North Island, Varroa mite has decimated the feral population, threatening crop production. He advocates breeding a bee that is tolerant to Varroa so that the feral population can be restocked. Some progress has been made, but up until now the resulting bees have not been as productive as traditional honey bees (*Otago Daily Times*, 6 August 2009).

Bee life cycles

Bees lead complicated lives. Their life cycle starts with the egg. The queen bee lays fertilised eggs in queen cells and worker cells, and unfertilised eggs in drone cells. They are vertical when laid. The egg falls gradually onto its side as the worker bee larva develops and in three days the larva hatches and is fed 'bee bread' by nurse bees. The larva goes through five instars or stages and four moults, checked and fed constantly by the nurses. After six days the cell is capped by the workers and the larva spins a cocoon. From this pupa an adult bee emerges in eight or 10 days.

The development of the queen bee is slightly different in that she is fed more royal jelly, secreted by the nurse bees, and for longer before the queen cell is covered over with wax. Also, the composition of the royal jelly fed to a queen changes slightly as the larva develops and hence it is not just the amount of royal jelly fed to the queen larva that causes it to develop into a queen.

Bee colonies in winter may number only a few thousand, but they build up numbers for the summer foraging season. If you can count 200 bees a minute going into a hive, it is thought that the colony will number about 30,000.

There are three types of bee in the hive: the queen, the drones and the workers, all designed differently to fulfil their separate roles.

The queen is the only fully developed female in the hive and she is the mother. Her job is to lay eggs in the brood cells, up to 2000 a day during spring. She produces pheromones, chemical emissions used by animals and even plants to send messages. Queen bee pheromones keep the worker bees sterile and stop the development of new queens. While she can produce them, the queen remains in control of the hive. When the queen begins to lose vigour or the colony prepares to swarm, the workers create queen cells and have the queen lay eggs into them. A new queen emerges; the first to do so often stings the other queen cells to remove the opposition. If more than one queen emerges, they fight to the death and the winner takes the hive. The old queen either departs with a swarm of about half the bees in the colony, or dies. There is only room for one queen.

The queen is fairly easy to identify, being bigger than the others. She has shorter wings and a large, tapering abdomen. She does not have pollen-gathering structures on her legs or wax-producing glands on the abdomen. She is a specialised layer of eggs.

The drones hatch from unfertilised eggs, in larger cells than the eggs for the worker bees. They are the males and have no sting. Their purpose is to mate with the new queen. She leaves the hive and high in the air, to ensure that only the fittest drones reach her, she mates with as many drones as she can. This is to collect enough sperm to last for her lifetime. When she is running out of sperm, the worker bees sense this and start the replacement process. Their work done, the drones die after mating.

Worker bees are all females and they are the most numerous. The worker has large wings, a long tongue and a barbed sting with which to defend the hive. Pollen is collected on the body hairs and is swept by means of combs on the hind legs into a pollen sac.

Worker bees have rosters, depending on their age. They start off as cleaners, cleaning cells and looking after eggs, then proceed to feeding the larvae and nurture and feed the queen. After this they receive honey and pollen from the foragers. The next stage is the production line, wax-making and cell-building. A little later, they become entrance guards and then their final job is to go out foraging for nectar and pollen.

Research has identified a substance called Vitellogenin, a protein that has multiple functions in animals, including keeping them young. Mature bees are fed less of this substance and this may influence division of labour and foraging behaviour; the lack of it also shortens their life span. It is said that a worker bee visits over 4000 flowers to make a tablespoon of honey. All this activity, as well as the change of diet wears them out quickly, and the life span of a summer worker bee is only about six weeks, half of which time will be spent foraging. In the autumn the bees live longer, with the help of more Vitellogenin, and some of them will overwinter in the hive (R Oliver, *American Bee Journal*, August 2007).

Domestic arrangements

A beehive is an artificial nest and the natural nesting site for bees is a hollow tree. Hives are normally rectangular boxes with eight to 10 parallel frames, which hold the vertical honeycomb. This is where the eggs, larvae, pupae and food are stored. The brood nest spans the inside frames and the two outside combs at each side are usually used for storing surplus honey and pollen.

Pollen collected from plants is protein food for developing larvae and newly hatched bees while honey provides the energy, or carbohydrates. After feeding themselves on honey and pollen the nurse bees produce 'royal jelly' which is fed to all larvae for the first three days of life, and thereafter only to queens. After three days, workers' and drones' diet changes to worker jelly or 'bee bread', a mixture of honey and pollen. The queen is fed pollen all her life since she needs the protein to make eggs ready for laying.

If a queen excluder is used (see chapter 4), surplus honey will be stored in 'supers' above the brood chamber, and this allows the beekeeper to harvest the surplus without upsetting the colony or the brood nest. If you take honey that is needed by the colony for winter food you will have to replace it with sugar or corn syrup.

When they open the hive, beekeepers look for signs of a healthy community, showing all is well in the bee world:
- A brood box containing eggs, larvae and sealed brood cells
- Surrounding cells full of food: bee bread (moistened pollen and honey), pollen, and honey
- Surplus honey in the supers
- A queen in the brood box (if she is not found, evidence of her presence in the form of brood is enough).

There is preparation to do, of course, as with any livestock, before you introduce the hive to your chosen site. Bee colonies consist of thousands of insects but they function as one individual and this individual must be catered for. Will there be suitable flowering plants available for your bees and is there likely to be a succession through the year? Good bee plants

Bees drinking in the author's garden

have plenty of nectar and also pollen, needed in large quantities when larvae are being reared. Wattles are a good source of pollen, as you will know if you suffer from pollen allergy. Hobby beekeepers with one or two hives can usually allow their bees to stay in one place and find enough food, which will avoid the need to move hives.

Is there plenty of water? Bees need a great deal of water for their activities but it must be accessible; if the container is too deep they may drown. They need a landing place, a ledge or some robust leaves on the surface of the water. They alight on a firm surface and drink from a damp surface. If there is no natural water source near your proposed site you will need to provide a container of clean water that the bees can access easily. If you don't, they may end up drinking from the neighbours' swimming pool, which will not be popular.

Will bright lights from the street or the neighbours be visible from the hive at night? If so, the hive may need to be shaded from the light source, although this may not be a problem. Bees only forage when plants are secreting nectar, so they are normally only flying by day. If they are disturbed at night they will go to a light, but otherwise they won't worry about it.

If you think that bees will cross neighbouring gardens to get in and out from their hive to the food sources, it will be necessary to erect a barrier. You can't fence bees in, but hedges or a trellis 2–4m high will force them to gain altitude and keep out of the neighbours' way. Warmer months, when people are more likely to be out in their gardens, are also the peak activity time for bees.

Swarming

Swarming is the natural way in which colonies multiply and of course it is one way to start in beekeeping or to expand your stock. Once you are an established beekeeper you may be contacted and asked to remove swarms from places where they are a hazard and if you have a spare hive this will be a way to expand your enterprise. However, usually you do not want your bees to swarm as it depletes the colony of workers.

The reasons why bees swarm are several; in general, they are not satisfied with conditions at home, 90% of the time due to overcrowding. It's as well to be observant and to take steps to keep the colony happy if you don't want your bees to disappear over the horizon in a black cloud. Swarming is influenced by the queen.

Swarming is less likely to occur with young queens than with old ones, which is one reason why queens are replaced.

The queen may not be producing enough pheromones to control all the workers – often an older queen.

If the weather breaks during a honey flow the bees will give less food to the queen, which may decrease her egg laying and also her production of pheromones. The workers will no longer be inhibited from building queen cells. Bad weather in summer can cause bees to swarm.

Congestion in the hive, when the beekeeper has not given the bees enough room to store honey, will mean that the queen's effect might not be felt throughout the hive. The trick is to give them more room when it is needed, that is when there is a good honey flow and the bees are very busy. If they have nowhere else, honey will be stored in the brood compartment and this reduces valuable brood rearing space. But it is not a good idea to give them wax foundation, sheets of wax embossed with cell pattern, if it can't be used.

If there is a strong colony, a honey flow and the bees are not comb building, they are likely to swarm.

Communication

As we have seen, the queen communicates with the workers by means of pheromones, and odours can guide bees to suitable plants. Bees also have a dance language, which was studied extensively by Karl von Frisch. Bees coming back to the hive perform a series of movements on the comb. The orientation of the dance is thought to correlate to the relative position of the sun to the food source, that is, its direction. The length of the 'waggle' part of the dance is correlated to the distance of the food from the hive. The more vigorous the dance, the better the nectar or pollen supply.

There are two schools of thought about the bee dance and much controversy surrounds them. Upholders of the dance theory agree that odour may also be involved but their opponents insist that odour alone is communicated and that the dance has no meaning at all.

It has also been argued that the bees see each other on the same level, whereas the human observers have watched the dance and seen the pattern from above. Everything else bees do is done for a purpose, so would they waste energy on a dance that had no practical use? There is still a little mystery about bees.

CHAPTER 3

Getting Your Bees

His labour is a chant
His idleness a tune;
O, for a bee's experience
Of clovers and of noon.

EMILY DICKINSON

There are various ways to acquire a hive of bees. You can start from scratch with a nucleus colony, bought from breeders. It will contain a queen and enough workers to get the colony going. A local beekeepers' association may introduce you to people who sell 'nukes'. Spring or early summer will be the best time, when the queen is laying the maximum number of eggs and there is plenty of nectar and pollen about to encourage the breeding up of more workers.

If you buy a nucleus they will be supplied in a smaller box and be transferred to the new hive you have waiting for them. There are two ways in which the transfer can be done. The bees can be smoked (see below) and their frames moved into their new hive, and any stragglers moved gently along with a soft brush. Or, your beekeeping mentor may suggest that the two boxes are placed side by side and the bees are let out for a few days to orientate themselves before you move their home to one side and place your hive in the same position. Foraging bees should then return to their new home. Using a smoker, you then lift out the nucleus frames one at a time with any bees on them and transfer to the new hive. Shake any remaining bees off at the new hive entrance and fill all the remaining space in the new hive with frames of beeswax foundation for them to work on.

Alternatively, you can take over a working hive from a beekeeper and relocate it, again when food is plentiful. This may be an easier way to start.

Before buying, check for bee health and soundness of the hive structure. The bees and their hive will be moved to your location, often with the entrance open, loaded on to a trailer at night and left a while to let the bees settle down. (Most movements in beekeeping are unhurried, with frequent pauses.) The transport and arrival are all done during the hours of darkness. It is best to avoid having to stop at a fuel station and it is preferable not to go through brightly lit streets, unless the hive is closed. Artificial lights will attract bees.

Moving bees with the hive entrance closed is also common, particularly for small operators. It makes for ease of handling, allowing you to travel safely and unload without stings. The entrance is closed at night when most bees will be in the hive; bees at the entrance can be driven inside by the smoker. Ventilation will be needed or the bees will suffocate during the journey. To provide ventilation, a hole can be made in the side or bottom of the hive and covered with insect netting to keep the bees in. Alternatively, the lid can be completely removed and the top of the hive covered with insect netting nailed down to the super.

It is wise to take an experienced beekeeper with you to inspect a colony before purchase – the bee equivalent of kicking tyres at a car yard. But do some observations yourself, based on your research. Dead or dying bees about the hive entrance will be a warning, even to a beginner, that something is amiss. So will an unpleasant smell when the hive is opened. Other signs of disease may need a more expert eye cast over the frames. Watch the flight path of arrivals and departures. Do the workers fly straight and vigorously or are they uncertain and erratic? Colony strength may be roughly estimated by the number of arrivals per minute, if you are visiting on a warm, still day when most bees are foraging.

The colony you buy should have a young queen, which is busy laying eggs; if it hasn't, you should be able to buy a young queen separately to get off to a good start.

Taking a swarm is another way of starting up a new colony, but swarms are unpredictable and you can't be sure there will be one handy when your hive is ready.

Experienced beekeepers arrange a 'baited hive' to attract stray swarms, the bait being an empty hive with a faint odour of bees, perhaps provided

The bees can be smoked ...
and their frames moved ...
(Nic Moore)

A drip tray, made by Ron Branch, to put under a hive to avoid making a mess when transporting it.

by an old, empty brood comb which has previously been used by bees. (More about swarming above.) This is a cheaper way to start and it might be achieved with the help of an experienced operator, but it's a matter of luck with regard to timing and to the behaviour and health of the swarm. You will be keeping bees of unknown pedigree, but if they turn out to be aggressive this can be bred out of them by introducing a docile queen.

Beekeepers and the law

Of course there are laws relating to the keeping of any livestock and bees are no exception. The rules are intended to protect the health of all bees and as I mentioned before, bee health is of great concern. Check the legislation for your state to ensure you comply with the requirements. The first rule is registration as a beekeeper, even if you have only one hive. Registration is with your state Department of Primary Industries. States differ slightly but the aim is the same: to know who is keeping bees and where. In Queensland, registration costs an annual fee of $10. The intention is not just to part beekeepers from their money, but to be able to control disease. On registration you will be issued with a registration number and this must appear on the side of every hive; it is suggested that you burn the number into the wood to make it permanent. Registration also entitles the beekeeper to financial compensation if hives become infected with some diseases and, in some states, free testing for American foulbrood spores in the honey.

The number of hives you are allowed to keep may be restricted. In Queensland, if you live in an urban zone you are not allowed to keep bees on a block less than 400 square metres, but between 400 and 1000 square metres you can have two hives. Rural areas will have different restrictions and should be easier, but it would be prudent to check.

You must have removable frames in the brood box, which means you will need to buy or make a standard hive. This is a legal requirement to enable government inspectors to remove frames to look for diseased bees. The old-fashioned skep, although looking really good, is illegal because an inspector would not be able to inspect the skep without first destroying it.

Notifiable pests and diseases must be reported to the Department if you suspect that your bees are afflicted. In NSW, Victoria and Queensland these include:

- American foulbrood disease
- European foulbrood disease
- Small hive beetle

You may also be required to keep a record every time a hive is transported; this includes place of origin, destination, date and operator. It applies mainly to commercial apiarists who move to chase the nectar flow or travel long distances for pollination, but if you do move a hive for any reason, record it.

Checklist

Before your bees arrive you should … finish reading this book! And then do the following (for suppliers, see Further Information at the end of the book):

1. Familiarise yourself with bees, consult a mentor, and attend field days and training courses if available.

2. Register as a beekeeper with your state department of Primary Industries, and obtain their beekeeping code of practice and a copy of the relevant regulations.

3. Choose a site for the hive or hives: it should be fenced off from livestock, in a quiet and secluded spot, with winter sun and summer shade. Easy access will be needed to mow the grass, add new boxes or frames, or remove the honey harvest, so the fence should have a gate wide enough to admit a trailer or trolley.

4. Prepare the site: mow the grass, remove tall weeds, and erect a screen if you need to deflect bee flights upwards. Prepare a base for the hive if the ground is uneven, with a stand if there are cane toads about.

5. Provide a water supply, with runways for bees to land.

6. Obtain a hive kit and assemble the hive; OR construct your hive to exact specifications (the bees expect very precise measurements); OR purchase a new or used hive.

7. Put the hive in place – sterilised and repainted if second-hand, with your registration number painted on its side, and install the frames with a

wax foundation (unless you bought a complete outfit, in which case the hive will come with the bees).

8. Purchase protective clothing: one set for each person who will be involved.

9. Assemble your basic equipment: a smoker and fuel, a hive tool, spare frames and the wax foundation.

10. Have a bee sting treatment ready for use. Some people use antihistamine creams. Calamine lotion is soothing. Caladryl cream contains calamine and an analgesic, and is said to work well. According to one intrepid reporter who was stung several times in the cause of research, the home-spun remedies toothpaste and ice are the best pain relievers.

CHAPTER 4

Housing and Equipment

How skilfully she builds her Cell!
How neat she spreads the Wax!
And labours hard to store it well
With the sweet Food she makes.

Isaac Walton

It's always good to consider how domesticated livestock lived in the wild, or how their wild cousins live today. This gives us many clues about their preferences and some of their prejudices too. Wild bees nest in dark places, in hollow trees or between rocks that afford protection from predators and also from extremes of temperature. They like to be near accessible water and as close as possible to a range of flowering plants. It is also worth remembering that any dark furry object reminds them of bears; that's why beekeepers wear smooth, light clothing.

Whether they are in a hollow tree or in the latest desirable, white-painted bee residence, honey bees always do the same thing. They build a nest or 'comb' made up of six-sided wax cells, the wax secreted from glands on their bodies. Some combs are filled with honey and some with pollen. Brood combs are the nurseries, containing the larvae, the developing next generation.

The hive must be sited carefully because there are many things to consider, some of which have already been touched on. We know that for reasons of safety, bees must be as far as possible from paths, roads, parks and schools or be placed near a fence or barrier to deflect the flight path upwards on their leaving the hive.

It is most important to provide them with a suitable water supply from the start. Once they have found an unsuitable one, such as a nearby swimming pool, they will still go there even if you give them water near the hive. Bees are creatures of habit and they need to be trained in good habits, so make sure that they always have plenty of water at hand. Water is needed to help to cool down the hive in hot weather, as well as being used for diluting honey to feed to the brood. There should be a creek or a dam within 500 m of the hive, with rocks or ledges for easy access. If your site has no handy water source you will need to provide clean water at all times.

From the insects' point of view, they like to be secluded. People may interfere with the hive and even in rural areas there can be hazards. Beehives should be fenced off from farm animals and at least 3 km from commercial crop operations unless they are organic and bee friendly. Ants and other bees can harass your bees and cause them to go on the defensive, which means stinging.

Remembering that honey bees originated in warm climates, we can work out that the colony is more comfortable with the hive in a warm and sunny position if the temperature is below 20°C, as it often is in the southern states of Australia and in New Zealand. Extremes either way cause the bees much more work and stress. Above 30°C they need shade of some sort, preferably under trees, and this may mean that the hive has to be moved according to the season. White hives are traditional and they reflect heat. Worker bees work hard to control temperature and humidity in the hive; extreme conditions cause them stress and also divert them from honey production. We can help them by facing the hive entrance away from prevailing winds, especially cold winds, but if the entrance can face the morning sun this may encourage them to get up early.

Housing

Hive stand

It is a good idea to make a base for the hive. This will keep the immediate area clear of vegetation, will help ventilation in hot weather, and keep the flight path clear. Wooden bases may warp or be attacked by white ants. Many home beekeepers believe that concrete bases are the best option.

... here the hive rests on a brick and timber stand (Jen Owens) ...

This hive stand is made of galvanised piping (Nic Moore) ...

... whilst these hives sit on concrete.

They will keep the hive plumb and won't sink into the ground when the supers are laden with honey, or if the ground becomes soft in wet weather. Hives leaning at an angle look very unprofessional!

The hive

If you buy a hive as a complete unit you will need to check that the box itself is sound. Pine and cedar are often used. Plastic boxes have the advantage that they don't need preservative or paint and they are becoming more common.

The entrance is at the bottom of a hive (Jen Owens)

A typical hive is a rectangular wooden or plastic box, raised up a little from the ground. Some hives in Queensland are placed on a stand to keep them clear of cane toads and the stand may also offer protection from ants and some forms of bee parasite.

A bottom board, made of galvanised iron, tempered hardboard or timber, supports the hive and in it there is an entrance for the bees, which has a door or 'closer' for fastening them inside when you want to move them. On three sides of the bottom board there are risers of 9–20 mm to allow room for the bees under the frames.

The brood box is the bottom storey in the bee house, as we saw before, with the brood comb slung between frames. This is where the queen lays her eggs and the workers tend the emerging young. The brood comb contains larvae and pupae and stores of pollen and honey. On the top there is usually a wire or plastic grid with narrow slits. This is the queen excluder, designed to keep her out of the upper hive so that no eggs are laid there. Neither drones nor the queen can squeeze through the slits and eggs won't be laid in the honey super, but the smaller workers can move about freely.

At Api Mondia 2007 in Melbourne (the biennial world apiarist convention), an American researcher gave a talk on beehive design. He said that there had been little research into hive design due to the conservative nature of the industry. His research showed that there was no difference as far as the bee was concerned, and honey production, brood laying and swarming were not affected by hive design. He did note, however, that a large hole made in the hive base, and covered with netting to allow air to circulate, caused the queen to lay many more eggs compared with a hive with little air circulation.

Above the brood box are the supers, additional boxes of eight or 10 frames hanging vertically, of a standard size to fit the length of the hive, which is 500 mm (21 inches). The width of the hive will vary according to how many frames it can hold.

An eight-frame box is 347 mm and a 10-frame box is 400 mm. The frames are moveable and can be taken out to extract the honey; moveable frames are now required by law in Australia. On top is a lid or hive cover.

The frames are in four pieces. The top bar has a groove along the bottom side; there are two end bars, each with four holes through which to thread wires, and a bottom bar. Traditionally made of wood, frames are now available in plastic.

This super has eight frames (Nic Moore)

Bees at work in the frames (Nic Moore)

Each frame has four strands of bee frame wire, which is threaded as a single length through the holes in the end bars, and then tension is applied. The wires support the beeswax foundation. Young bees produce wax from glands under the abdomen, hanging motionless in clusters while the little scales of wax appear from their glands. They scrape off the scales with their mouth-parts and shape it into comb. The six-sided honeycomb shape is the most efficient that could be used, as it uses the least material to store the greatest amount of honey. The hexagonal structure has been copied in modern engineering.

Because of the effort involved in wax making, beekeepers often supply their bees with ready-made wax foundation obtained from bee suppliers, so that the insects have more time to make honey. Sometimes they use a starter strip about 5 cm wide across the top of the frame and then leave the rest to the bees. Full sheets of foundation, however, cause the combs to be straight and even. They are easier to remove to check on disease, and also to harvest the honey.

Bees are fussy. If the cell size in the ready-made foundation is not exact, the bees will tear it down and use the wax to build their own comb. They like fresh foundation, so you are advised to buy it only when you need it.

Beekeepers use an embedding tool to attach the beeswax foundation to the wires in the frames. When you buy foundation, check that it is made from pure beeswax, with no adulteration such as added paraffin wax. Plastic foundation is quite acceptable to bees.

Hive sizes are standardised partly because, as we saw in the history sec-tion, Rev. Langstroth designed an improved moveable comb hive in 1851. He based it on the earlier discovery by Huber that there is an optimum 'bee space' of 10 mm between the parts. If the space is greater or smaller than this, the bees

Bees are fussy ... The cell size has to be correct. (Jen Owens)

will change it or will not use it. Langstroth lived in Philadelphia, USA, and his success led to the adoption of the Langstroth hive as standard in many parts of the world. This has the added advantage that hive parts are interchangeable and easily obtained.

If you are a keen carpenter and decide to make your own hive, stick to the recommended designs and the bees will be happy. The wood preservative you use, usually copper naphthenate, should not contain any pesticide.

The precise details of how to build a beehive are contained in the NSW Department of Primary Industries publication, *Bee Agskills* (2007), although they advise beginners to use factory-made equipment.

When the honey is harvested you will need somewhere to process it. An extra shed will add to the expense, but you may be able to use a toolshed or garage, provided it is clean and, most importantly, it can be kept bee-proof. If you use your garage to extract honey, make sure the door is completely shut when you start or all the bees in the neighbourhood will come to visit you at once.

If you decide to sell honey the processing room will be subject to food hygiene regulations and permits (talk to your local government health department).

Equipment

As in any other field, do some research before you buy equipment. If you buy second-hand equipment you need to know current prices; websites, catalogues and beekeeping magazines should give you some idea. The disease status is also important. Second-hand material may harbour disease, which could affect your bees.

One internet site offers a new beginner's outfit including hive and equipment for $553 (2009 prices).

The smoker

Puffs of smoke are traditionally used to quieten the bees when you work with them. As a beekeeping friend of mine said, 'Don't overdo the smoker.' Think of each bee with its compound eyes watering! He also suggests, for your own sake, using a fuel that gives off a pleasant smell, such as pine needles or stringy bark. The theory is that the hint of smoke makes the bees think the forest is on fire and so they eat honey in preparation for leaving,

The smoker is an ingenious yet simple piece of equipment. (Nic Moore)

which calms them down. Another possibility is that the scent of smoke interferes with the bees' reception of the queen's scent messages and masks any alarm pheromones that guard bees release when an intruder invades the colony. Some people suggest that smoke intoxicates them and makes them drowsy. Whatever the reason, it is universally used to calm bees. But too much can flavour the honey and smoked honey is not a gourmet's delight. Particles of soot are not wanted.

The smoker is a metal barrel, which holds smouldering fuel and has a nozzle in the lid. Bellows attached blow air into the bottom of the barrel and force smoke out through the nozzle. Obviously, care must be taken: the smoker can be a fire risk in careless hands. It is good practice to have water at hand to deal with unwanted flare-ups.

The smoker fuel can be anything of vegetable origin: dried pine needles, papery bark from trees, wood shavings or wood chips (providing the wood hasn't been treated with chemicals), or dry leaves. The smoke should be fairly dense and cool. Bees don't like the smell of burning chemicals, hair, feathers, or rags containing oil. One beekeeper I heard of used a firelighter in his smoker and the bees went absolutely berserk.

To light the smoker, start it off with small amounts of crumpled newspaper, then puff the bellows gently and add small amounts of fuel. Stir it up with the hive tool (see below) and when it is properly alight, fill up the barrel with fuel. You should be able to put the smoker down and leave it alight – it may be needed again. Bob Owen suggests using a barbecue lighter to start the smoker as the long snout will allow you to start the fire more easily from the bottom of the smoker.

A smoker can be extinguished by blocking the nozzle with a piece of screwed up paper. This stops airflow through the smoker and so the burning ceases. Alternatively it can be filled with water. After use, the equipment should be washed in soapy water.

The hive tool

Shaped like a small jemmy, this is used to prise the top bars of honeycomb loose, and for scraping. A screwdriver is often used as a hive tool because it has the strength to prise up a frame that is stuck. Most Australian beekeepers use the so-called Australian J-tool to inspect hives; the handy J-type end is used to prise up frames. An alternative design is the American Hive Tool. This is not often used in Australia, since prising up frames is more difficult with this tool.

The smoker and hive tool at work

The hive tool and protective clothing are essential equipment for any beekeeper. (Nic Moore)

Protective clothing

You will probably feel more confident at first if you go for the full protective gear, although with experience you may operate with less. Before you climb into the suit, get rid of watchbands, jewellery, wool and dark clothing. Bees do not like hair gel, after-shave lotion or anything strong smelling; we used to be advised to wash with scentless soap. Try not to approach them while exhaling fumes of onion or beer!

Handling should be slow, patient and cautious at all times, gentle and not jerky. You should try not to squash any bees when you work with them. The odour of a squashed bee puts the others on the defensive because it gives off alarm pheromones. That warns the other bees of danger and they will attack you.

You will need a hat with a broad brim to keep the veil over your face. However, it should stand clear of the face, because if it is near the bees might sting you through it – and they tend to go for faces. The bee veil will be black, either wire gauze, which is the safest, or cotton netting, which tends to be more comfortable. Use black netting because it causes much less glare than white, and you will be able to see what you are doing much more clearly.

The conventional wear is white, one-piece overalls with long sleeves, made of a smooth material. Some may have a white hood attached instead of the hat, over which you can wear a wire face-piece. White and light material will avoid overheating in warm weather. Make sure that the overalls are baggy, both to allow circulation in hot weather – beekeeping is hard work – and also to allow free movement when you bend over to look in a hive.

Non-wool socks and heavy-duty boots with steel safety caps are recommended, with protective gaiters round the ankles. The boots will protect your feet if you move the hive. Wrists can be protected by tightly fitting armlets, tubes of thick cloth with elastic at each end. Leather gloves will complete the outfit. And now we are ready to bring on the bees.

Bill Ringin, dressed in protective clothing for removing the frames of honey from the hives.

CHAPTER 5

Food for Bees

Where the bee sucks, there suck I;
In the cowslip's bell I lie;
There I couch when owls do cry.

WILLIAM SHAKESPEARE

Nectar is the raw material from which bees make honey. It contains sugar and is produced by glands on plants called nectaries, which regulate the sap pressure in the plant as a sort of valve. Nectar produced varies in quantity and quality, affected by the weather, the time of day and also by the species of plant. It provides energy for bees, the energy needed for growth, breeding, flying and keeping warm.

Pollen is the main source of protein for bees and they need pollen with at least 20% protein. The protein content varies a great deal according to the species of plant that produces it. For example, spotted gum pollen is 25–33% protein, whereas pine tree pollen contains only 5–7% (see www. honeybee.com.au for detailed information).

Bees can draw on body protein when they are stressed and they also use it to make royal jelly to feed to the brood. Strong bees that live a long time have high body protein (over 60% crude protein). This is important in autumn so that they are strong and disease-resistant over winter. Low protein bees can have less than 30% protein and they will be poor producers with susceptibility to disease. Body protein is reduced by extremes of weather, honey production, wax production and an increase in breeding.

Stress levels for bees will alter their protein needs and they only increase body protein at times of low stress. As we saw before, bees used for crop pollination need to be strong.

A heavy nectar flow means a lot of work for bees; all those flowers to visit, all those loads to carry back to the hive, all that work inside the hive for the younger bees. So does an increased breeding program, which may be stimulated by a strong nectar flow. Temperature extremes will stress them, just as they do us, and even some plants are more stressful than others. White Box, (*Eucalyptus albens*) flowers in winter. It has pale bark and waxy flower buds and fruit. It causes stress because the weather is cold, but the honey flow is good and so the bees are stimulated to breed. It all becomes too much for them, especially as the pollen protein in this case is rather low. Sometimes in these circumstances they produce a great deal of honey and then the population collapses.

Conversely, lucerne (alfalfa), a food crop for livestock, flowers when the weather is often hot and dry, so it stresses bees with too much to collect in the heat. The amino-acids in the pollen protein have been found to be low in some cases and, to make matters worse, when the anther (the pollen-bearing part of the flower) is tripped it kicks the bee, which seems most unfair. They must hate lucerne.

The best stress management for the bees described above is a natural one: a supply of better pollen with more digestible protein. Spotted Gum (*Eucalyptus maculata*) is an excellent pollen source for bees and also produces nectar for energy, except in severe drought, so the bees get a balanced diet. This tree occurs in Queensland, New South Wales and parts of Victoria.

Good food sources will depend on where you live. They include:
- Broad-leaved Stringy Bark (*Eucalyptus calignosa*)
- Faba beans (*Viccia faba*) – one variety is the broad bean. They are a winter crop, and have been grown commercially since the 1980s for livestock and human food.
- Tea Tree (e.g. *Melaleuca quinquenervia*) down the east coast
- Blakelys Red Gum (*Eucalyptus blakelyi*) in Victoria, NSW, and Queensland.

Conditions in Australia vary so much that the practical thing to do is to consult local beekeepers about the plants that are useful to them. In Western Australia, for example, apiarists will tell you that they rely on forest stands of Karri, Jarrah, Banksia and shrubby desert species. Tasmanians rave about Leatherwood honey, produced from Leatherwood, *Eucryphia lucida*, the most important native plant for bees in that state. The tree grows in the wetter west of the state as an understorey species beneath taller trees. In Victoria there is an almost mystic reverence for Red Gum (*E. camaldulensis*), both for honey and for firewood. Victorian bees also rely on Yellow Box, Round Leaf Box, Tea Tree, Ironbark and Stringybark.

Depending on times of flowering, bees will usually be able to access more than one species and this will help to overcome any shortfalls in nectar and pollen.

In urban environments, garden plants are important to bees, either as the main menu or as the background to main food sources. Later we will look at growing gardens with bees in mind. It's surprising that many weeds are good food sources; capeweed, dandelion, thistle and cat's ear (flatweed, or false dandelion) are visited.

The various species of eucalypt provide 80% of the commercial honey harvest in Australia, with weeds providing a significant part of the remainder. However, Bob Owen points out that our introduced bees have developed both their food needs and their food gathering techniques with European flowering plants in mind. Urban and semi-urban areas of Australia, with their abundance of gardens containing European flowering plants and weeds, are ideal locations in which to keep a beehive. Many urban beekeepers keep healthy thriving hives on top of a garage in the inner city. Rural Australia, where native plants predominate, is often not that bee friendly, since bees often have difficulty in removing the pollen and nectar from these plants. Bob Owen says it is the large number of gum trees in rural Australia that make beekeeping there attractive, not the ability of the bees to obtain food from these plants.

Unfortunately, flowering in eucalypts is difficult to forecast as it depends on rainfall. I have noticed our trees produce fresh leaves and buds after good rain, but then hang on without flowering for months if the rain dries up. There are also cycles over several years for various species, which experienced beekeepers can predict.

CSIRO has developed an information system specifically about eucalypts. It's called EUCLID and it details 894 eucalypt species with a bewildering array of facts.

There are various books and publications to help you. But whatever reference source you use, flowering times can only be approximate. It will

Flying pollinators like bees form their first impression of plants from a distance. They are attracted by colour and smell, colour being more important as the distance increases. Bees are supposed to be attracted mainly by yellow, blue, purple and ultraviolet colours. They cannot see red as humans do, so red flowers have to rely mainly on their scent to attract bees.

be a good idea to observe your bees and work out what and where they are collecting, then keep a dated record that will be useful for the future.

When you watch the bees, note how long they stay on a blossom: the longer the visit, the better the yield. If you can identify the tree species, look it up to see how long it is likely to flower; that will help to predict the nectar flow. If the bees are coming in heavy laden and landing short of the hive, the flow is really on and in this case you may need to take away surplus honey and add another box to give them more room.

Feeding the colony

There will probably be times when your bees will need help with nutrition, nectar for energy and pollen for protein. This may sometimes be done by moving the hive to another source of flowers.

Species like Ironbark and Yellow Box are not good pollen producers, and the lack of protein may cause the hive to decline in population and, of course, in bee longevity. The bees can then be moved to a site that will help to build them up again with good supplies of pollen. Commercial beekeepers now know the analysis of nectar and pollen for many plant species, which helps them to anticipate trouble. A colony can be fed extra pollen or pollen supplement, to avoid the adverse effects of poor or no pollen in their diet.

Supplementary feeding

Old-fashioned beekeepers may tell you that ideally bees should live on honey and pollen all the time and should only be fed in an emergency, such as a bad drought when they can't gather enough food to survive. Hobby beekeepers do not rely on bees for their living and so are unlikely to take the hard commercial view that it is good to remove honey and replace it with sugar, which is much cheaper. Bees, like humans, are better on natural food. So the beekeeper who works with nature will only remove surplus honey – which in some years means a very small or non-existent harvest.

Bob Owen tells the story of a five-star hotel in India that served a honey for breakfast that was absolutely tasteless. 'It was really obvious that the bees in that hive had not seen a flower for some time and that the honey was made out of sugar only. The moral of this is that bees need a mixture of nectar and sugar to survive the winter,' he says.

However, feeding may be needed if the hive is short of food and moving is not practicable; winter conditions can create this situation. The bees may be fed supplements to make up the shortfall of whatever nutritional components they lack, or honey substitutes for their whole diet when there is a complete lack of nectar or pollen. Protein can be stored in the form of pollen for use later; some beekeepers collect pollen for their bees.

Carbohydrates are usually fed in the form of sugar syrup when the colony is short of honey, say in the winter, or needs stimulation. The best nectar substitute is white cane sugar, sucrose. It is fed dry or in syrup form. If you make it into syrup, don't burn it; caramelised sugar is bad for bees. Corn syrup is often fed in North America. But avoid brown sugar, which can give the bees dysentery. Dark, unrefined sugars like molasses, treacle and golden syrup are lethal to bees.

In an emergency dry sugar can be fed, half to one kilogram in a heap on the inner mat of a hive, without interfering with the colony. Dry sugar is frowned upon in some circles because it gives the bees so much extra work to do. They are used to handling nectar, which is a dilute sugar solution and so they dilute the dry sugar from their own body water – and then proceed to evaporate it as usual. Then they have to rush off in search of a drink. So a liquid sugar solution is much more natural and they don't have to leave the hive to look for water.

Sugar will stimulate the bees into getting on with their job, if they are lacking food. It will encourage them with breeding and subsequent foraging for pollen (which is why bees kept for pollination are often fed sugar); it

Sugar-syrup feeder made by Ron Branch. Several litres of sugar syrup are poured into it, and the box – which is the same size as the hive box – is fitted just below the lid. The bees can drink from the holes without drowning.

will energise the hive. Beekeepers use it to help the bees through out-of-season honey flows. It will also prepare bees for pollination, start them breeding earlier in the spring and can be used when you wish to prepare for queen breeding. But as Bill Ringin, an experienced beekeeper says, 'Sugar syrup and so on is a good short-term fix but, in the longer term, bees do better on honey.'

Honey from another source is sometimes fed to bees, although sugar syrup seems to help them more. If honey has candied in the combs, this can be kept and fed to the hive as needed. There is a disease risk with honey and you will need to know the source of any honey you feed.

In spring, or when queen rearing, the colony will need to be stimulated. *AgNote DAI/178* from NSW Agriculture recommends small quantities (1–2 litres) every few days of a 1:1 concentration of sugar and water by volume.

To provide stores for winter, feed the colony in autumn with a weekly dose of 5–10 litres of the syrup until they have enough honey stored. The winter syrup ration is thicker: 2 parts sugar to 1 part water, which makes a dense syrup. Syrup should never be thinner than a 1:1 ratio as the bees will have to work too hard to get the sugar. You can check the amount of stored honey in the hive by lifting it; if it is light, they will need sugar.

There are several types of sugar syrup feeder, the top feeder being the most usual – inverted buckets or jars with small holes in them placed over a hole in the lid of the hive. Another type of sugar solution dispenser is a flat tray the same size as the hive lid. It fits under the lid and the bees can drink through perforations in the metal cover. Do not leave fermented syrup in the feeders.

Feeding protein

Protein in the form of pollen may be needed at times. Some beekeepers take pollen stored in combs from the hive in times of surplus, keeping it to feed back to the bees later.

Pollen substitutes are used in some cases, although bees seem to prefer the real thing. Such foods need to be affordable and not toxic; high levels of things like oil or salt can kill bees. Naturally, the particle size of any feed must be very small. Soya flour, sorghum and triticale have all been used in mixtures. Brewer's yeast and baker's yeast are useful. Vitamins and minerals are sometimes added, but there is not very much evidence as to what the bees require.

When deciding whether to feed supplements you need to know whether the food is for maintenance or breeding (just like any other livestock). If you are building up numbers and have encouraged an expanding brood nest, the protein supplement will be needed until the bees can go out and get pollen for themselves.

Pollen substitute can be fed in powder form or in patties. Pollen itself can of course be fed on its own but it is usually extended with cheaper substances such as soya flour or yeast, with pollen at least 5% of the mixture. Pure icing sugar or white sugar makes the mixture more appealing to the bees and it can be up to half the mix, depending on requirements. The sugar to water ratio should be 70:30. Patties do not keep very well and should be stored in the freezer if they are not used immediately.

Sometimes feeding supplements don't work because the bees don't like them; more sugar and pollen in the mix should improve the uptake. You can now buy ready-made protein cakes for bees.

Honey can be used instead of sugar, and this makes the cakes softer and easier for the bees.

The bees in winter

Bee colonies can die out in winter. The time to ensure your bees' healthy survival is in autumn. Inspect them carefully in April, on a warm sunny day. Take off the lid and look at how much honey is stored; check the brood for symptoms of disease and inspect the queen. If the queen is laying drone eggs or has disappeared, the beekeeper will decide not to overwinter that hive, and it will be joined with another colony. When colonies are mixed, a sheet of newspaper is put between them. Over time the bees will chew through the paper, allowing the two groups to mingle gradually, which minimises stress. (With this kind of management in mind, you may decide that you need more than one hive of bees.)

If the colony is small it may not have enough bees to keep up the hive temperature and they may die of cold in the winter. If you have started with a nucleus hive in mid-summer it will need to have built up its numbers by autumn. Once again, small colonies can be joined – but one queen must go. Joining colonies should not be done after mid-autumn at the latest, as it will stress the bees.

In autumn, colonies are reduced to two boxes or even one, and they should have one box nearly full of honey. Always leave too much honey rather than too little, but try to keep the hive as compact as possible, so the colony can control the temperature more easily.

You may not see many bees in winter, but they are not asleep. 'Winter' is so variable across Australia and New Zealand that it is hard to generalise, but there will be a decline in activity during the colder months. The optimum outside temperature for bees in winter, when they consume the minimum food, is 10°C. In autumn, the workers feed the queen less and as a result she usually stops egg-laying for a while and the bees are comatose. But as the temperature falls they need to become more active and take in honey, to release the energy as heat to maintain warmth in the hive. Bees cluster in cold weather, and keep changing places from the outside to the inside of the cluster. (I have seen small birds do the same thing in an English winter, and penguins do likewise in the Antarctic.)

For winter, hives should be insulated and draught-proof. The entrance can be reduced in size to a few inches, to cut down draughts and make guarding easier. For best results the hive should be on a warm, sunny site, protected from prevailing winds.

Condensation in the hive may be a problem in cold areas because, when the bees are working hard to keep the place warm, they increase their rate of respiration. Good insulation will minimise condensation, and if the hive has a slight tilt towards the entrance, any moisture will tend to run out. One small company in Victoria, 'Amazing Bees', sells a hive that is designed to prevent condensation. It is made of pine, 19 mm thick. The roof consists of two plywood boards, separated by a 40-mm insulating air gap, and covered by a dark-painted tin lid. Vents in the air space provide air circulation to ensure that moisture is not trapped.

In warmer areas bees may forage all winter and continue to produce brood and in this case you will need to inspect more frequently.

If bees are fed lots of sugar in the winter this may encourage them to keep breeding and they may use up their stores and die of starvation. Feeding must be managed to work with nature.

Traditional beekeepers will tell you not to interfere with the bees in winter. If the hive is heavy enough – they need 11–16 kg of stored honey per hive – they can be left alone until the beginning of spring. Hives should not be opened until the weather has warmed up and the bees are on the move again. When you do make an inspection, don't leave the lid off too long, because the eggs need to be kept warm.

CHAPTER 6
A Bee Garden

To set budding more
And still more, later flowers for the bees,
Until they think warm days will never cease,
For Summer has o'er brimm'd their clammy cells.

JOHN KEATS

You could make, or adapt, a garden for your bees, with year-round sources of pollen and nectar. Tall shrubs, trellis or a hedge could separate them from the neighbours if necessary. Clumps of flowering plants make it easier for the bees to forage by cutting down their travel time and masses of colour can be part of your design. The other consideration is to provide a succession of blooms, from early spring until late autumn and through winter if possible. This is the objective of most gardeners and it's particularly important for the bees.

Although most commercial beekeepers in Australia and New Zealand rely on shrubs and trees – mainly eucalypts for their honey harvest – urban and suburban flowers keep bees very happy. They are Europeans and a mixture of native and European flowering plants are to be found in many of our gardens. They won't be too keen on the minimalist style of garden that displays a paved area and a couple of urns. Bees need something more natural than that; remember, though, that a bee's range is considerable, so they will forage over a large area.

Your first consideration is to make an organic garden, to do away with pesticides, herbicides and artificial fertilisers. A growing number of home gardeners are following organic principles and substituting compost and

A good bee garden will have lots of plants that flower when bees are most active, with a good mix of indigenous and European plants, including fruit and vegetables.

Bees love salvias: Salvia 'Blue Hills' left, with day lilies, yarrow, and taller-growing salvias in the background. (Penny Woodward)

chook manure for chemicals. The idea of feeding the soil instead of directly force-feeding the plants is catching on. If you haven't tried it, you may be surprised to find that your garden soil retains moisture better after a dressing of compost. The plants will be happier and so will the bees. Feed the plants well for a good yield of nectar and pollen. Plenty of nitrogen is needed for good quality pollen.

When plants are wilting for lack of water, they have a reduced nectar flow, so water is essential in the bee garden. Mulching will reduce evaporation from the soil and most gardeners use a variety of mulches. A rainwater tank will be a great asset and a grey-water system from the house will ensure that very little water is wasted.

Ideal garden varieties

The older, open varieties of plant are better for bees than the F1 hybrids, which are sterile and do not set seed. Flowers originally developed their colours and scents to attract insects but, even though the modern hybrids are often bigger and brighter, it's all show. They will yield very little, if any, pollen and nectar. The traditional plants are better for you, too, because you can save the seed from them and grow it the next year. Most gardeners enjoy saving seed to grow and to give to friends.

I must confess to a liking for the old-style cottage plants that are easy and cheap for us to grow and good for the bees. They can be grown from seed and cuttings gathered from friends' gardens – it needn't cost the earth to set up a bee garden. The bees prefer single flowers, such as the old varieties of rose, to the double-petalled ones; the new ones have less nectar and are harder to reach. All those frilly petals keep the bees away from the nectar.

Old-fashioned plants include marigold, poppy, aster, foxglove, geranium, sweet peas, lambs' ears, and older rose varieties. Sunflowers are attractive to bees and, when they set seed, they feed the birds as well. You can buy wildflower seeds in packets and the bees will appreciate them.

If you take off the dead heads of flowers, this will help the bees by encouraging the plant to produce more flowers (but only if you don't want it to produce seeds).

Your garden should have fruit trees if you have room: apples, pears, cherries and plums provide spring blossom, and they depend on the bees for pollination. Almond blossom makes for a bitter honey because the nectar is bitter, but it mixes in with other nectars. Ornamental flowering cherries are pretty and practical for bees.

Bees love aromatic plants, and your bee garden should include the herbs the settlers brought with them to be of use in the new world: lavender, rosemary, thyme and sage, including the ornamental salvia. They like roses, buddleias and lilacs. In fact, it should be possible to manage your garden to provide flowers in every month of the year. To do this, keep a record of what flowers when, and how long it lasts. Seasons will no doubt vary and summer drought can delay the autumn flowers but, in time, you will build up a picture of what grows in your particular area.

Here is a list of some of the bee plants found on our farm and garden (including vegetables and weeds – some weeds have several uses):

- Acacia (Australian acacias produce nectar from glands rather than the flowers – our local sticky wattles (*A howittii*) are sticky for this reason
- Alyssum, a tiny, sweet-smelling ground cover
- Banksia
- Berberis
- Borage, and its cousin comfrey
- Bottlebrush
- Broad bean in early spring, dwarf and runner beans in summer
- Capeweed
- Chickweed
- Clovers
- Daisies of various kinds
- Dandelion
- Eucalypts: Blue Gum, Grey Gum, etc.
- *Euphorbia peplus* (petty spurge, the cancer weed)
- Flatweed (cat's ear), a good source of pollen

- Fruit trees, including apple, pear, nectarine, orange and cumquat
- Geranium
- Grevillea
- Herbs: lavender, marjoram, mints, thyme, rosemary, sage
- Lambs' ears (the leaves are said to be a cure for bee stings)
- Loquat tree, which flowers profusely
- Melissa or lemon balm is sometimes called bee balm (but the name properly belongs to several plants of the genus *Monarda*). Melissa is a member of the mint family and has a sweet lemony scent; bees visit the white flowers.
- Nasturtium (under the fruit trees)
- Nettle
- Roses.

Weeds for bees

The definition of a weed is 'a plant growing in the wrong place' and gardeners take pride in their weed-free gardens. But many weeds are visited by bees and if you do decide to give up all chemicals, there will be some weeds about. Many of our garden weeds are of European origin. They travelled here with people in various ways and were not found in the wild. Such plants, now designated as weeds in Australia, were once cultivated, or they were gathered for a wide variety of uses. You might find it possible to enjoy your weeds.

Weeds in farmland also play their part and many farmers in England have special flower meadows, which are not cut for hay or silage until the wildflowers have bloomed and set seed. (Modern farming practice decrees that the grass is cut much earlier, and from fields where no 'weeds' are allowed.) These meadows help to provide food for bees and also to ensure that rare wildflowers survive. There is a move in the USA to encourage farmers to plant small patches of bee plants in odd corners; borage, lavender and daisies have been suggested.

Chickweed   Also called starweed, chickweed is used to make a healing ointment. The leaves are small and tender and it makes good compost. It can be used instead of cress in a salad.

Dandelion   The true dandelion, with big indented leaves and one flower per stem, is very useful to bees. In many parts of the world the dandelion is the first bee flower of spring and beleaguered beekeepers in North

America are asking people not to mow their lawns until the dandelions have flowered. Dandelion is also valuable to herbalists for its medicinal properties. The leaves are eaten in salads, the big yellow flowers are used to make wine and the roots are dried and ground to make dandelion coffee. So you can easily justify a bed of dandelions in your garden; the drawback, and the reason they are on the weeds list, is that the seeds, like little parachutes, are spread by the wind and so the plants can pop up anywhere in the garden.

Goosefoot or fat hen The young leaves are used as a vegetable and prehistoric people ate the seeds. The flowers are greenish and not very conspicuous, but bees seem to like them.

Nettle There are several nettle varieties, including the dead nettle with bigger flowers and no sting in the leaves. This is a plant with very many uses; stinging nettles are a nutritious vegetable and their fibre was once important for making cloth. Nettles indicate a rich soil and there should be a place for them in a bee garden.

The food garden

Many fruit plants have flowers that bees love. Bees are used as pollinators on commercial crops of citrus fruit, although the nectar production is variable. They are used on mangoes, avocados, stone fruit and melons. Bees are important for good yields from kiwi fruit. Raspberry flowers have plenty of nectar and pollen, the nectar with a high sucrose content, and bees carry out 90% of the pollination on raspberries.

Bees are essential in growing some vegetables and if you grow your own vegies, the bees will be there to help. Honey bees are said to be poor pollinators of tomatoes, but little native bees can pollinate the little cherry tomatoes.

Sweet peppers do self-pollinate, but bee intervention will help. Pumpkins and peas and beans need to set seeds, but leaf vegetables are eaten before they flower.

The current interest in home-produced food is encouraging more backyard beekeeping and the bees play an important part in our food ecosystem. Even in years when the honey yield is low, the beneficial effects of pollination will improve the yield of fruit and vegetables in the garden.

CHAPTER 7

Managing Your Bees

/Merrily, merrily shall I live now
Under the blossom that hangs on the bough.

William Shakespeare

Spring

Spring is a moveable feast. We have early springs and late springs, cold springs and warm springs, so it's easier to go by the feel of the air than the calendar. But by late August the bees should be at the front of your mind again, with particular reference to food. In late August and early September, more bees die of starvation than at any other time of year, because they eat up their stores during winter. The danger is that when the weather is warmer the bees start again to rear brood, and if the weather then reverts to cold and wet, as can happen, the food will be used up quickly to feed an expanding brood nest and maintain temperature. Bees operate their premises at a warm 35°C within a range of a few degrees.

The field bees at the entrance to the hive in spring will give you some idea of what is going on inside. If more pollen is being gathered as the weather warms, this will indicate more brood rearing, since pollen is food for the brood and this in turn means that the queen is doing her job.

Remember, however keen you are to open up in spring, observe the precautions.

Pick a warm day with no wind, in the middle of the day when most of the bees are out foraging. Whenever manipulating the components of the

Be careful what fuel you use i[n]
the smoker. (Nic Moore)

Use the bellows gently
to fan the flame once lit.
(Nic Moore)

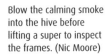

Blow the calming smoke
into the hive before
lifting a super to inspect
the frames. (Nic Moore)

Above left: Prising up a frame with the hive tool. (Nic Moore)

Above right: Inspecting the frames. Note the protective clothing. (Nic Moore)

Left: A burr comb (Nic Moore)

hive, try to be as efficient and smooth as possible and avoid injuring or crushing bees – they will hate you for it.

Give the guards at the entrance a whiff of smoke from the smoker; then raise the cover very slightly and give two or three puffs of smoke over the frames before taking off the cover. If the bees start to get excited they can be given another puff or two.

To take out a frame, prise the adjoining ones apart so that there is clearance, otherwise bees may be crushed. The comb that bees sometimes build outside the frames is called a burr comb and it may have to be cut through near the adjoining frame. This might be a delicate operation if the comb contains honey. Take out the frame very gently and hold it upright, or unsealed cells may drip honey.

Your first concern is the health of the colony and you need to make a careful inspection for disease (see chapter 8). Spring is the time when disease is more likely to surface, since the overwintered bees are older, and they may also have consumed infected honey during the winter. There may be evidence of disease in some combs and not in others, so each one must be inspected.

The next thing you do is look for the queen. From August to about the end of May there is likely to be, under normal conditions, brood in the hive. Some people mark the queen with a spot of paint on the thorax to make her more visible. The brood nest is the obvious place to look for her. Remove one of the brood nest's outer frames and place it in a spare box, then inspect the frames one at a time, leaving a one-frame gap between the ones you have inspected and the rest. If there is brood in two boxes, look first in the bottom box. If the queen is not there or on one of the combs you will need to take all the combs out and check the floor and walls of the hive.

The brood nest should contain eggs, larvae and sealed brood. If there is no brood at all, it's no use searching for the queen. In a good light you should be able to see the tiny eggs, or small white grubs, in the bottoms of the cells.

If there are eggs but no larvae or sealed brood, this may mean that the queen has died and the eggs laid by a new young queen. Sealed brood and queen cells will mean that the colony is preparing to raise another queen.

Re-queening

The queen is located and assessed by her activity. She may be a few years old, or your bees may be aggressive and you want to introduce a quieter strain. Whatever the reason, if you need to re-queen, spring, when the bees are collecting supplies of nectar, is a good time to do it. You can buy a new queen by mail or email; a commercial breeder will post her to you in a small cage and with an escort of a few workers. At one end of the cage is a plug of 'queen candy', a mix of honey and icing sugar.

The queen can be kept in her cage for a few days but it is best to get her into the hive as soon as possible. The old queen is killed once the new queen has arrived safely, but before she is introduced to the hive. There is only one queen in a hive!

There will probably be a cork sealing each end of the travelling cage. Remove the cork in the end with the candy and put the queen cage between two frames of brood in the middle of the brood nest, with the candy end

Bill Ringin requeening.

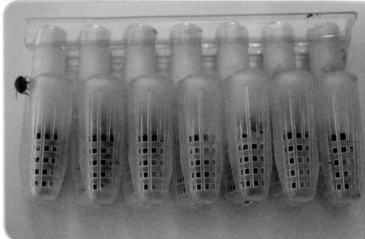

Queens in their travelling cages

tilted slightly upwards. This is so that if an escort worker bee dies she will slide back down the cage and not block the exit. You then close up the hive again and leave it alone for at least a week, which allows time for the new queen to be accepted by the colony. The bees will consume the candy plug and release the queen.

When you open the hive again, look for eggs. That will mean that the queen has taken up her duties and all is well. An added benefit is that a colony led by a young queen is less likely to swarm.

If the colony is strong, or when there is congestion, you may want to split it. Another hive will be needed, of course. At this time of the year you may also want to be prepared in case you are able to catch a swarm; already we have two reasons for adding to your number of hives. As we saw in chapter 2, bees are likely to swarm if there is a strong colony, a honey flow and they are not comb building.

Honey can be taken in the spring when conditions are right, and this will provide fresh combs for storage of honey and reduce crowding; it may also prevent swarming. If there is plenty of nectar and pollen available for the bees and the weather is suitable, neither too hot nor too cold, there will be surplus honey and the bees have the summer in front of them.

However, spring will mean an increased use of stores and the food supply must not be allowed to get too low. If that happens the bees may cut down on brood production. If you have several hives, honey stores can be equalised between them, but only if you are sure they are free from disease. In case of trouble later – such as summer drought – it will be well to leave some honey.

If you remove full combs and replace them with empty or extracted combs, this will encourage the bees to produce more honey, so long as there is plenty of nectar available.

Weak colonies

Some colonies will emerge from winter with a drop in population. They can be united with another colony, or helped to get through and build up numbers on their own, which you will need to do if you have only one hive. You can help them to conserve heat in the hive by taking away supers that are not being used. If there is space in the brood chamber, a division board may be placed to keep the area compact.

Extra feeding will stimulate brood rearing and help a small colony to build up faster (see chapter 5, Food for Bees).

Spring is a good time to remove unsuitable combs while the population is relatively low. Damaged combs or old ones with misshaped cells, or those with too many drone cells, should be taken out. It will be wise to melt them down straight away so that wax moths don't find them. New sheets of wax foundation can be used to replace them.

Summer

Beekeepers never interfere with the hive unnecessarily, but it is as well to keep an eye on the status of the colony. If the colony is getting smaller, inspect for disease and check the queen and the brood pattern she is producing.

Most beekeepers replace two or three old combs each year in summer, if they have not already done so in spring. Keep checking for wax moth in any stored combs and small hive beetle in combs containing honey.

You are now hoping for surplus honey. The nectar flow will dictate conditions in the hive and it is by no means predictable. This is a case where local conditions prevail and should be closely observed; your bees will be responding to their own environment. You will have studied the flowering plants in your area within 2 km of the hive, and have some idea of when flowers are expected, although of course plants vary, depending on sun and rain. Looking at the number of unopened flowers will give you some idea of how long the flowering period will last. It is a good idea to keep a diary and record the plants that give you nectar flow each year; there may be a pattern that will help you in the future.

Heavily laden bees will sometimes land short of the hive entrance and this is a good sign. Look at the bees coming in (you need good eyes for this) and check the pollen on their legs. If there are several colours of pollen this means that the bees are visiting several different kinds of flower.

While commercial honey producers often go for 'single line' honey such as yellow box or clover, this involves ensuring that the honey is taken while nectar is coming from only one source. Hobby beekeepers usually prefer whatever mixture the bees can find and this may give them a better chance to get the pollen and nectar they need for their diet.

When you open the hive, look for plenty of stored pollen round the brood combs and a line of new white wax along the top of the comb. Later in the day you may see bees fanning with their wings at the entrance, a sign of great activity inside.

If you lift the hive using the back hand-hole, the weight will tell you if there is surplus honey. Open up and see how much honey has been capped with wax. Remove frames from the super and if two-thirds or more of the cells are sealed, honey is ready to be taken.

The honey should be mature before it is harvested. To check, hold an uncapped comb with honey in it on its side and give it a shake. If loose nectar drips out it is not yet ripe; the bees have not yet reduced its moisture to the right level.

On the other hand it is possible to delay too long before taking honey. If you leave it until too late in the autumn the bees may have moved some honey out of the supers into the brood area, as the amount of brood falls. The white wax cappings might be darkened by the bees' dirty feet if they are left too long.

There are several ways of taking the combs of honey from the hive and the first objective is to separate bees from honey, to remove the insects from the frames you want to harvest.

The traditional way to get rid of bees is the 'shake and brush' method. On a warm still day, gently puff smoke over the tops of the combs to drive as many bees as possible down into the hive. Take out a frame and shake it quickly to get rid of the bees, either down into the box or at the hive entrance. Brush off the remaining bees with a clean soft bee brush, available from any beekeeping supplier. It is not to be used for anything else, and is

You can shake the bees from a super before removing a frame. (Nic Moore)

Removing a frame for honey extraction. (Nic Moore)

better than a small kitchen brush, since the narrow row of bristles will minimise the chance of bees getting caught in the bristles. Horsehair bee brushes are better than nylon ones since they stay clean longer.

Put the combs in an empty box and keep them covered to prevent robbing or repopulating by the bees. Take the honey away immediately; the whole operation should be done neatly and quickly, with no pauses. (Those who don't favour this method say that it is the easiest way to upset the bees and get stung and that it might not be a good method for urban beekeepers with near neighbours. But it has the merit of being cheap and quick.)

Another method involves using 'escape boards' and leaving them in place at least overnight. This might be the best option for beekeepers in areas where the temperature goes down at night, because it assumes that the bees will leave the honey supers to cluster for warmth and maintain a favourable temperature for brood below. You can buy up to 20 different types of escape or 'clearer' boards and the time taken to clear the bees

An escape or clearer board helps to empty a hive of bees – they can exit easily through the funnel exits in the corners, but the bees have trouble finding their way back in through the small openings. (Nic Moore)

effectively will vary. The board has several one-way exits that allow the bees to go down into the hive below, but not to come back up. There should be an empty or partly empty super below into which they will go.

The top super should be the one that is ready to be harvested. Take it off and place the escape board (making sure it is the correct way up) under it, then replace as before. There should be no gaps that might allow the bees to get back into the honey super. Leave overnight as a rule, although some beekeepers only give the bees an hour or so to make their escape. The next morning, remove the honey super and brush any remaining bees off the frames.

Beekeepers with large numbers of hives use blowers to remove the bees, and some people think this is the best method. Bees are less likely to respond aggressively to a strong wind than to a bee brush, but a bee blower with a petrol engine costs nearly $600 and also uses fossil fuel.

You may read in old books of the use of chemicals to remove the bees, but this is not acceptable these days; we don't want contaminated honey.

Water again

In summer it is particularly important to check that the bees have enough water at all times and as we saw before, this should be established before the hives are placed on site. If their water supply dries up or is covered in algae, they will go further afield to find water and this will reduce their capacity to make honey. It may cause them to seek moisture from a pool or garden water feature, creating a problem with neighbours. If a water supply fails them, the bees may not return to it when it is re-established.

Some bees specialise in water carting, collecting about 25 mg at a time and making 50 trips a day. We can make it easier for them. In hot weather the colony will need several litres of water a day, for controlling the temperature and humidity in the hive, as we have seen. On a hot summer day the wax combs would melt if the bees did not control the internal temperature (and we can also help them with shade from the sun if needed). Water in the nest evaporates and keeps the hive cool. And water is needed for feeding the developing bees and for diluting food.

The best source of water is clean, flowing water in a stream or creek, accessible from leaves or ledges; but many people will need to provide water for their bees. Small dishes of water don't work, as bees find the water source by the increased humidity of the air above it. Therefore you need a large container with mesh netting across the trough to keep other animals out. The bees will need landing rafts on the surface so they don't drown; wood, cork, branches and leaves will make a platform for them. Bees don't generally land in or on water but on damp surfaces, and absorb moisture with their mouth-parts. The water must be pure, for obvious reasons.

In hot weather some people provide water inside the hive, using a syrup feeder arranged so that the bees don't drown. The main point to remember is that whatever arrangements you make for watering your bees, it has to be checked frequently and particularly during days when the temperature is high. It's worth repeating that it is most important that the water supply is reliable and does not dry up.

Autumn

The frames should be inspected at the start of autumn and also at the end. A hive will need six frames full of honey to see them through the winter. It sometimes happens that the autumn flow is disappointing and your bees are left with less than this, so you know that they will need supplementary feeding (see chapter 5).

It is also time to inspect the queen again to see that she is still active. If the areas of eggs she lays are compact that is a good sign; if they are scattered, she may be losing vigour. You may be able to see that she is damaged in some way. You can re-queen in autumn if you need to.

Healthy brood has a solid pattern with not more than 10% of the cells empty. If there are more, disease may be present (see the section on health).

In autumn the bees may have a renewed burst of activity as the flowering plants revive from the summer heat. That's why it is important to check at the end of this period; you may be pleasantly surprised at the honey store. I have noticed in our area that autumn is often a very good time for bees and they are high up in the eucalypts or down among the salvias and lavenders after we have had refreshing rain.

Winter space

It makes sense to give the bees only the space they need for winter. If they have more it will be harder to keep the hive warm, so any supers not covered by bees should be removed and they may be reduced to one box for winter.

A sunny spot facing north or north-east will be best for winter, but with the entrance away from prevailing winds. If the ground is likely to be damp in your chosen spot the hive can be placed on a stand, so long as it is secure and not likely to tip over.

Bees are left alone in winter and apart from checking the food supply on a warm still day, you will not need to do anything if they have enough food. This is the time to repair and refurbish equipment ready for the spring. You may want to make or acquire another hive, or get some more equipment. This is a good time to repair frames and paint hive components. Any paint smells should have dissipated by the time they are needed in spring.

A winter hive with only one super and very little activity

CHAPTER 8
Diseases, Pests and Parasites

*Honey is sweet,
but the bee stings.*

PROVERB

This chapter is intended to help you to look out for common bee problems, but you should always seek professional advice. The Department of Primary Industries in each State and Territory is the best source of information as to which diseases are currently notifiable and within what time period. They will also advise on proper identification of the problem and possible remedies. It is very important for the industry as a whole that all beekeepers are registered and that any notifiable disease is reported promptly.

Talk to beekeepers about their experiences and you will develop knowledge of what to expect in your local conditions, but make sure the person you talk to has sufficient knowledge of diseases and how to diagnose them in the hive.

There is much concern in rural circles about the health of bees and the various threats to the health of colonies, from both endemic and exotic diseases. For this reason bees are only allowed to be imported into Australia under strict conditions.

The Australian Department of Agriculture, Fisheries and Forestry (DAFF) points out that the economic loss resulting from the loss of bees is not confined to honey producers. Bees are vital for the pollination of major crops: almonds, avocados, cotton, cherries and other stone fruits, apples and pears, melons, kiwi fruit, berries, pumpkins and seed crops such as clover and lucerne all depend on bees for production.

Just like any other livestock, bees can be affected by a variety of ills. Some bacteria, viruses, fungal diseases and parasites can invade the hive and the careful beekeeper will always be on the watch (with your glasses on if needed!) for the first signs of anything that appears to be abnormal. The careful observation of a colony will be essential; every time you open up the hive you will be making mental notes, and you will then be much more likely to spot trouble as soon as it arises.

Form a picture of what a normal colony looks like. Healthy brood has a regular cell pattern with no dead larvae or pupae. The caps are slightly convex and brown, tan or cream, without holes. Healthy larvae are shiny and a pearly white with an orange line down their backs and healthy pupa darken from white, the eyes showing colour first.

Diseases

Stringing of brood killed by American foulbrood (AFB). (Bill Ringin)

American foulbrood

The main bacterial diseases are American foulbrood and European foulbrood.

American foulbrood (*Paenibacillus larvae*) or AFB is the more deadly of the two because the bacteria form spores that have been shown to remain viable for over 35 years. This is worldwide and destructive and in Australia and New Zealand it is a notifiable disease. If you know or suspect the disease is present you must notify an apiary inspector (State DPI) within 24 hours. When you report the disease the hives are inspected and samples are usually taken to obtain a specialist laboratory diagnosis.

AFB is spread by spores of the bacteria which get into the food of larvae up to three days old. They germinate in the larva's gut and grow there, penetrating the nearby tissues. The larva normally dies after the cell has been sealed; the bacteria also die but leave thousands of spores behind. Adult bees are not affected, but they carry the spores and the hive is gradually weakened because there are fewer larvae developing into adult bees.

The caps of cells containing infected individuals may be sunken, dark and sometimes perforated as the bees have partly removed the caps. The dead larvae and pupae will be lying on their backs on the lower cell walls. If you stir their decaying remains with a matchstick, they will be ropy – able to be drawn out like a rope – and there will sometimes (but not always) be a characteristic smell.

Later, the decaying larvae dry to a hard, black scale in the cell, which is impossible to remove without damaging the cell walls. You can imagine how the disease spreads; worker bees clean out the infected cells and spread the spores through the colony. Honey stored in the infected cells will contain spores. If the colony is weakened by the disease, robber bees may take the honey back to their hives and spread the infection. Beekeepers can spread it by moving frames or supers from contaminated hives to healthy ones.

In mainland Australia it is illegal to treat AFB disease with antibiotics. This is because research has shown that the use of antibiotics will suppress signs of the disease, but that the spores will survive and can be unwittingly spread by the beekeeper to healthy hives. In New Zealand it is also illegal to use antibiotics or drugs in beehives.

Once the disease is confirmed you will be advised of the necessary steps to take. The requirement varies and in some areas the whole infected hive has to be burned. In other places the frames and combs are burned and the empty boxes and sound hive components may be sterilised. This may be achieved by gamma radiation or hot dipping in paraffin wax, microcrystalline wax or a combination of both. The fact that the spores can live for over 30 years is a good argument for buying new equipment when you start beekeeping. It is best to avoid second-hand material that has not been in use for some time and where its history is unknown. If you do buy previously used material, take steps to sterilise it before use.

In New Zealand, there is a national pest management strategy and one of its aims is to eliminate AFB in New Zealand. It's thought that this is possible as incidence of the disease is low and the Varroa mite is eliminating feral bees that could spread the disease. Beekeepers are getting rid of AFB in individual enterprises by destroying infected colonies and preventing the spread of the disease to other hives.

European foulbrood

European foulbrood (*Melissococcus plutonius*) or EFB is also a notifiable disease. EFB is caused by a bacterium that also lives in the gut of larvae, but this one does not form spores. In this case the larvae are usually affected before they have straightened and before their cell has been sealed. EFB can spread quickly through a hive. Dead larvae are curled or twisted in abnormal positions in their cells, become off-white and darken and are soft and watery. The diseased larvae dry to form scales, which unlike AFB can be easily removed from the cells. In some cases, other bacteria may invade larvae killed by the EFB bacterium and when this happens the larvae can be somewhat ropy when the matchstick test is used. In these cases it is sometimes difficult to distinguish whether these larvae are infected by EFB or AFB and a laboratory diagnosis may be needed.

It is thought that vigorous healthy colonies can survive this disease and that it affects colonies under stress. The disease seems to be more prevalent in Victoria than in other Australian states.

EFB can be controlled with an antibiotic, oxytetracycline hyodrochloride, where it is legal to do so, but this is not really desirable since honey from a treated colony may contain antibiotic residues. The antibiotic must be prescribed by a veterinarian.

Chalkbrood

Chalkbrood (*Ascosphaera apis*) is a notifiable fungal disease.

This fungus infests the gut of the larvae and competes with the larva for food. When it dies the fungus goes on to consume the body, making it look white and chalky (hence the name). When the remains dry and shrink they may be white, dark blue-grey or black. The dried chalk-like remains are called 'mummies'. It happens most often during a wet spring. Increased hive ventilation and keeping the hive warm is the best treatment. It is a good idea to replace the queen with a young, prolific, egg-laying queen, of a strain bred by a queen breeder.

Stonebrood

Stonebrood (*Aspergillus fumigatus*, *A. flavus* and *A. niger*) is another fungal disease. It does not seem to be a problem in Victoria.

These fungi are common in the soil and some Aspergillus species cause a disease in humans, a lung infection called Aspergillosis or Farmer's Lung.

The spores of the fungus lodge in the gut of the larva and form a ring round the head like a collar. The larvae become mummified and turn black. The larvae are covered with spores and the workers clean out the infected cells. A healthy colony may recover and this could depend on their level of hygiene. Some worker bees are more hygienic than others and in Britain, a scientist at Sussex University is trying to breed from hygienic strains of bee. British beekeepers are seriously worried after losing 33% of their bee population and have asked the government for funding to support more research. It is thought that improved hive hygiene would reduce disease risks for bees and it would be much better than suppressing disease with chemicals.

Sacbrood virus (SBV)

This is also a disease of larvae, affecting them just before they change into pupae.

A sac of liquid forms around the larva, underneath the outer layer of skin, which turns grey or pale yellow and then begins to dry and darken, the head darkening first. Dead larvae are stretched out with the head raised to the top of the cell at the cell opening, like a banana or gondola shape. Finally the remains are reduced to a flattened scale, darkening with age and easy to remove from the cell.

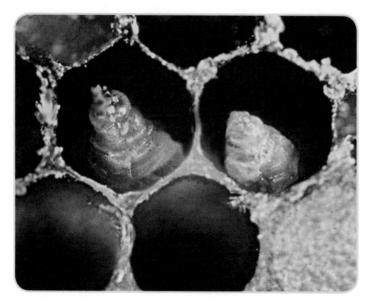

Sacbrood virus (SBV)

Israeli acute paralysis virus (IAPV)

This virus is the second to gain a lot of attention. It has been associated by US scientists with the Colony Collapse Disorder (CCD) mentioned previously. It sets bees' wings shivering and eventually causes paralysis. The question is whether it is a cause or a consequence of CCD. Healthy, imported Australian bees were found by US researchers to carry the virus, which was identified by genetic sampling. But it was later found by US scientists in samples of US bees, set aside some years before importation of Australian bees began.

Some scientists believe that symptoms of IAPV do not show up on Australian bees because they don't carry the Varroa mite (see below), which weakens them. There is much controversy about the whole subject. This just adds one more strand to the already complicated theories of bee diseases.

Pests and parasites

Nosema

Nosema disease is caused by two protozoan parasites, *Nosema apis* and *N. ceranae*, that invade the stomach of adult bees. Both these parasites form spores as part of their life cycle. Nosema is the most widespread adult bee disease in the world. *N. apis* has been around for years, but in recent years *N. ceranae* has been found in European honey bees in many countries including Australia. It's been detected in bees in Queensland, NSW and Victoria. Nosema is not easy to identify and poor performance of the colony may be blamed on other factors, but it can halve the length of a bee's life. *N. apis* is thought to be present in all colonies, only becoming a nuisance when conditions favour the parasite.

The effects of *Nosema apis* are noticed when adult bees die off early and young bees go out foraging in their place. The glands of nurse bees become unable to produce royal jelly and these bees may cease brood rearing and feeding larvae. If the queen becomes heavily infected she will stop laying and eventually die. Dysentery may be associated with outbreaks of Nosema, but not in every case. In bad cases, the whole colony may die.

A proper diagnosis can only be made in a laboratory, but symptoms to look for include a reduction in the number of bees, as mentioned before. There may be dead bees at the hive entrance (although often there aren't), and adult bees that can't fly (although this may also be an indication of another disease problem or the effect of pesticides).

In some cases, dysentery may be a symptom, and the hive may be covered in spots of faecal matter. Bad weather can increase the problem, as bees normally excrete their waste outside the hive when possible. As the weather improves and becomes warmer after the cold of late autumn, winter and spring, Nosema becomes less of a problem because the bees are able to excrete their waste outside the hive. The number of Nosema spores inside the hive decreases and the infection declines.

Beekeepers are still learning about *N. ceranae* as it is a relatively new disease. Scientists have found that, unlike *N. apis*, *N. ceranae* can cause losses of bees during the warmer months of the year.

Control   Chemical control of Nosema is not legal in Australia. *Nosema ceranae* has not been found in New Zealand.

Some of the good management practices already mentioned will help to reduce the effects of Nosema (and other diseases as well). Bees hate wet and cold winds, so the north side of a hill will give them shelter. They need as much sun as possible, except in the summer months. The hive needs to be dry. Beekeepers who put their hives on stands to deter cane toads have found a reduction in Nosema levels. Reducing the number of boxes for the winter and avoiding moving brood combs after the hives have been prepared for winter will also help. A strong colony with a young queen, a colony that is not too stressed by rearing brood in winter, will stand the best chance against Nosema. Low protein can also lower resistance to disease, and the colony will benefit from high-protein pollen in autumn and winter, if they have the choice.

Heating may be used to de-activate Nosema spores in dry material. Dry equipment is heated to 49°C for 24 hours to destroy the spores. Higher temperatures will damage the combs.

Exotic pests and parasites

These are seen as a serious potential problem in Australia and a risk to the entire beekeeping industry. All beekeepers are asked to keep a look-out for evidence of their presence, and to ask for professional advice if they think they have found anything suspicious.

Varroa mite

Varroa destructor and *Varroa jacobsoni* are mites that are present in parts of New Zealand but not, at the time of writing, in Australia, and both beekeepers and government departments are extremely keen to keep them out. The mites are about the size of a pinhead, red or brown, and shaped like a scallop shell. They can be seen with the naked eye on the thorax or abdomen of an infested adult bee, if you can spot a creature 1.1 mm long. But they spend much of their time inside brood cells, multiplying and feeding on pupae.

These mites used to be a fairly harmless parasite of the Asian bee but, in the past few decades, they have adapted to infest the European honey bee. An infestation will build up slowly but in time it results in deformed bees, shorter life spans and eventually the death of the colony. The mites feed off the body fluids (blood) of larvae, pupae and adult bees and, as if this weren't bad enough, they act as carriers for a number of viruses, the effect of one of which leaves the bee with deformed wings.

Testing for Varroa mite   It is suggested that all beekeepers should do two tests for Varroa mite at least twice a year, both in brood cells and on adult bees. For each test, the hive is opened with a smoker in the normal way.

The mites seem to like drone cells on the edge of the brood nest, so the beekeeper lifts out individual drone pupae with tweezers and examines them for the reddish-brown mites. Look inside the cell, as they may be lurking at the bottom. If there are no drone cells, examine worker pupae and cells.

The other test involves finding the mites on adult bees and a method has been developed that detaches the mites from the bees without harming the

Varroa mite

bees. Icing sugar causes the mites to lose their grip on the bee's body and they fall off.

About 300 adult bees from three brood combs are shaken onto newspaper and then poured into a large jar with a plastic or metal lid with small holes drilled into it and a heaped tablespoon of icing sugar in the bottom. Put the lid on the jar once the bees are inside and gently turn the jar so

1

A female mite enters the cell of a 5-day-old larva (preferably a drone cell) when ready to lay eggs.

2

The adult female mite hides in the jelly and waits for the cell to be capped.

3

After she has finished the brood food she feeds on the bee larva itself.

7

The mites attach themselves to adult bees, feed on them until they're ready to lay eggs and start the cycle again.

Life cycle of the Varroa Mite

(Varroa jacobsoni)

4

Sixty hours after the brood is capped the female begins to lay eggs in batches at 30-hour intervals.

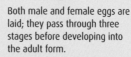

6

After mating, the male dies in the cell but the adult females emerge with the bee.

5

Both male and female eggs are laid; they pass through three stages before developing into the adult form.

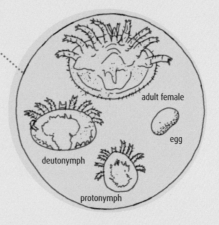

adult female

egg

deutonymph

protonymph

the bees become all sugar-coated. After a few minutes, turn them again.

Shake the icing sugar through the holes in the lid onto a sheet of white paper, in an area protected from the wind. Gently tip out the bees in front of the hive entrance and then look carefully for mites in the icing sugar, the tin and the lid.

What to do if you find Varroa mite (or any other exotic pest) If you are unfortunate enough to find any, put them in a small jar with methylated spirits and report the find to your nearest DPI apiary officer, but don't send samples off to them. Make sure you mark the hive for identification if you have several hives. Wash all tools and equipment and leave your protective clothing in a plastic bag at the site.

There is a small, brown pollen mite in Australia – smaller than Varroa – that is not a parasite of honey bees.

Honey bee Tracheal mite

Once again, this mite (*Acarapis woodi*) is a threat from outside which has not (yet) reached Australia.

These mites infest the respiratory passages of bees, so they are not visible externally. They were first found in the UK and have spread to Europe, the USA and South Africa. Female Tracheal mites, when mature, climb out onto a hair on the bee's body and then transfer to a young bee, where they move into the airway and lay eggs.

Tropilaelaps mite

This reddish mite (*Tropilaelaps clareae*), a parasite of brood cells, is also nationally notifiable in Australia. It either kills the brood or results in reduced longevity. Signs of this mite include deformed bees emerging from the pupae and deformed remains of pupae, both at the hive entrance.

Braula fly

Braula coeca is a wingless fly sometimes called a bee louse. It is found in Tasmania but not on mainland Australia. It is not a parasite although it lives on both queen and worker bees, as it feeds on nectar and pollen at the bee's mouth. If a queen has a number of Braula flies it may affect her capacity to lay eggs. The larvae of the fly make tunnels under the cappings of honeycombs and thereby ruin the appearance of comb honey sections.

Small hive beetle

This exotic (*Aethina tumida*) was first found in Australia in 2002 near Sydney and Brisbane, and attempts have been made to eradicate it ever since. Unfortunately it had spread too far and was even found in feral honey bee colonies (we can see now why feral bees are unpopular), so eradication was not possible and the pest was considered endemic.

Small hive beetle

The beetles are 5–7 mm, dark brown or black. They arrived in Florida some years before they were found in Australia and within two years of arriving there, the small hive beetle caused the loss of 20,000 commercial hives. So far in Australia, the warm, humid coastal fringe seems to favour the beetle.

The beetles can fly and can spread to other hives in this way, as well as on beehives and bees. The movement of soil around an infected hive could also spread the pest pupae.

Most, if not all, of these beetles pupate outside the hive, as the larva is attracted to light. The adults climb into the hive and the larvae infest the combs. The small hive beetle larvae feed on live brood and honey and they contaminate the honey, which makes it ferment and have an odour like decaying oranges. Small hive beetle may also infest combs taken from the hive and placed in the shed for extracting at a later time. In areas where beetles are a problem, beekeepers must extract the honey before it is ruined.

Looking for small hive beetle   The adult beetle is brown or black and up to 7 cm long. They avoid light where possible and try to find crevices or dark corners. Hives should be inspected for this pest. Take off the lid of the hive and put it on the ground upside down, then place the super on top of the lid for about a minute. When you lift it off quickly you may see beetles on the lid. If not, check the brood combs one by one and then the bottom and corners of the brood chamber. If you do find beetles, put them in a sealed jar and kill them by freezing.

Wax moths

And finally, we have a pest that does not attack bees but damages the valuable wax they produce.

Greater wax moth

Of the greater and lesser wax moths (*Galleria mellonella* and *Achroia grisella*), the greater wax moth causes the most damage. A pest that causes damage in stored combs, the wax moth flies and enters hives by night. It doesn't upset the bees themselves and wax moths are not thought to spread disease. The larvae of the moths eat combs, but not when they are covered with bees, so a normal, healthy, strong hive will not be troubled by this pest. They might damage combs in use, if the hive is weak for some other reason and the combs are not being used by the bees. They love combs that have had brood raised and/or pollen stored in them, though they can still damage other combs as well.

To guard against wax moths requires careful planning because they get into storage areas and lay eggs on stored combs and supers. Material and combs taken from the hive or the extracting room may already have wax moth eggs on them, and sooner or later these will hatch into larvae and damage good combs.

All stages of the moth can be killed at a temperature of minus 7°C for 4½–5 hours but a cool room will suspend all activity. It is necessary to begin timing the freezing process when the material and combs have reached this temperature.

As hobby beekeepers we may think our best defence is to try to seal the storage area (shed), especially when storing old combs there, which are particularly attractive to the wax moth. However, the wax moth may have laid eggs on the combs before we put them in storage and these will hatch over time and ruin the combs. It is best to follow Bob's method of control, as detailed below. In colder areas, winter storage in a light, unheated shed may do the trick. This usually works, but keep an eye on the combs because the greater wax moth can congregate and create their own heat, thereby raising the temperature of the microclimate and creating their own favourable environment – the cunning rascals!

You may wish to store a frame of honey over the winter or even for a few months in the summer, until you have time to extract. So it is, as we have seen, important to protect these frames against wax moth. Bob Owen says that a good way to do this is to seal the frame in a plastic bag and to place it in the freezer for two days to kill any wax moth larvae, pupae and eggs in the comb. It can now be removed from the freezer and kept in the garage or shed, still in its plastic bag to protect it against possible later egg laying by adult wax moths. Freezing may cause some varieties of honey to candy.

Wax moth larvae can also damage comb honey sections and bee-collected pollen, harvested for feedback to bees or for human consumption. Freezing of these products will kill all stages in the life cycle of wax moth.

Tutin honey, a New Zealand hazard

Tutu is a Maori name for plants of the Coriaria genus, poisonous shrubs or trees. If bees feed on the honeydew secreted by sap-sucking insects on these plants, the poison, tutin, a neurotoxin, can be transferred to the honey. In 2009 a beekeeper was fined for selling toxic honey. He did not know it was poisoned and had eaten some himself. The honey looks and tastes normal, and the poison's potency is not diminished by heat or time.

In January 2009, a tutin honey standard was set for New Zealand honey. The maximum level in extracted honey is 2 mg per kg and in comb honey, which is riskier, the level is 0.1 mg per kg. Honey harvested until the end of June must be tested if it is produced in an area with tutu bushes.

The main risk occurs in the Coromandel and Marlborough areas, between late December and the end of April. Management methods include taking honey before the risk period and watching the bees foraging. Poisoning is most likely to occur where there is a concentration of tutu plants with a high population of sap-sucking insects, dry conditions, and a lack of the kind of good nectar sources that would normally keep the bees away from the tutu.

Having considered the various pests and diseases that may afflict bees, we can see how important it is to be sure that hives, nucleus hives, and hive material are free from disease organisms and pests. It also reminds us to be extremely thorough and observant when inspecting a hive.

CHAPTER 9

Harvesting Honey

The pedigree of honey
Does not concern the bee;
A clover, any time to him
Is Aristocracy.

EMILY DICKINSON

A look at honey

Honey is a fascinating product, an ancient, natural source of sweetness. It cannot be imitated artificially; the bees add their own enzymes to complete the process.

You will have noticed how variable honey can be. The colour can vary from a pale, almost white colour, through hues of yellow and brown to nearly black. Some honey has a greenish or red tinge. When liquid, it is dense and semi-fluid, but sooner or later it will crystallise to be grainy, or sometimes buttery. Some honey solidifies so much that you can hardly manage to stick a knife into it. Honey varies a great deal in flavour and aroma and there are many different kinds to try.

The most obvious thing about honey is its stickiness. When you work with it, a good supply of hot water for cleaning up is needed.

If you want to liquefy crystallised honey, put the jar in water and warm it gently to 45°C, but not higher. High temperatures destroy the enzymes in honey and impair the flavour. Don't keep honey in the fridge, unless you want it to go solid.

Commercial honey

This is sold in several different forms:

- Natural honey is probably produced in your area by local beekeepers, and is filtered to get rid of its wax and pollen.
- Single species or mono-floral honey comes from one plant species, e.g. Yellow Box or Clover honey.
- Poly-floral honey comes from a variety of plants, and is sometimes called Wildflower honey.
- Blended honey may consist of honey from different plant species and even from different countries, and is usually blended by large producers and processors for both export and national distribution.
- Crystallised honey is honey that has set hard, which is quite a natural thing to happen over time, or that comes from specific plants (canola, for example) that tend to produce it.
- Creamed honey or whipped honey has been warmed, whipped, and then cooled to set smooth.
- Gourmet honey comes from areas where it has gourmet status, like Corsican honey, for example, which is certified in the tradition of French wines (*Appellation d'origine controlée*); many other countries have their own tradition of certification.
- Chunk honey is sold in the comb, straight from the hive; you chew the wax to get to the honey. In some supermarkets you can buy chunk honey in jars, with the combs suspended in honey.
- Organic honey has to be certified, and needs to be produced under certification by a recognised organic organisation. You would think that most honey was organic, and yours probably is, but it is difficult to prove.

Another type of honey

Honeydew honey is made when, instead of nectar direct from plants, bees collect honeydew, the secretions of sap-sucking insects such as aphids. It is strong and tends to be dark, and not so sweet as nectar honeys. This product of the hive is an important export for New Zealand. It mainly comes from the South Island, where the Red Beech and Black Beech forests support tribes of honeydew insects. It has a higher mineral content than nectar honey, due to the dark mould that grows on these trees, and also due to the extra influence of the sap sucker in the process.

The extra minerals and trace elements, measured by the electrical conductivity of honey, give it more nutritional value according to some sources. It also contains more complex sugars and a high level of antioxidants. On the down side, if the forests are wet and there are yeasts about, the honey can ferment. Honeydew honey is popular in Europe, and is produced in Germany's Black Forest or *Schwartzwald* (possibly with the help of their legendary elves).

Collecting honey from your bees

Like everything else in beekeeping, careful preparation is necessary before the great day comes when you harvest the honey. The space and equipment should be all scrubbed and prepared. Where are you going to work? Your home laundry may be the best place, if it has enough room and if you can keep bees, ants and other insects out of it. All windows and doors should be shut and gaps sealed in case your bees decide to reclaim their property, which can happen.

If you intend to sell your honey, take the right steps to get the necessary approval from your local authority for the work area and extraction plant. You might need to attend a food-handling course. It's a food and needs to be handled according to hygiene regulations although, compared with meat or dairy products, honey is easy to look after and keeps extremely well for years, even for centuries.

Equipment needed includes:
* sterile jars, or plastic 1-kg buckets with lids, to store the honey
* a plastic bin or bucket to catch the cappings
* a large knife
* a honey extractor, or a large piece of muslin
* a bucket to catch the honey
* a fine-mesh filter
* plenty of hot water.

Stainless steel, glass and food-grade plastic are used for honey equipment. *No copper, iron, steel, or zinc vessels should be used.* Honey is acidic and reacts with these metals, which can affect the flavour and colour of the honey and might even be toxic. Galvanised steel can be used for things like extractors, which are only in contact with the honey for short periods.

The work will be much easier if you have someone to help you.

The frames should be removed from the hive just before the honey is to be extracted. Waiting for the right conditions will have become a habit by now: a warm and windless day, with most of the workers out foraging. The frames can be cleared of bees by using the 'bee escape' overnight, or by using the smoker sparingly to persuade the bees to go down and away from the frames you want to take. Some people spray with water instead of smoke.

Keep the frames covered at all times because:

- As explained in earlier chapters, disease risks make it illegal under the Apiaries Act to leave the frames exposed to the air and the possibility of a visit by robber bees. This also applies to any equipment or tools with honey on them.
- Honey is hygroscopic, that is, it absorbs moisture. In a humid climate, or of course if rain is about, this can be a problem and there is a chance that the honey will ferment.

The task is to get the honey out of the comb and to separate the wax from the honey. Wax is also a valuable commodity and later we will look at using it.

Honey is thicker at low temperatures and the optimum temperature for extraction is 27–30°C. If you choose a warm day for the job this will make it easy; otherwise you may have to heat your extraction room. If, however, the room becomes hotter than 30°C, the wax combs will soften and might break.

First of all, the box or super is taken into the extraction room and the frames full of honey are lifted out. The bees have carefully sealed each cell of honey with a wax cap and this is now sliced off with a large knife. A carving knife will do the job; it will do it even better if it is heated by dipping it in hot water prior to use. Electrically heated capping knives are available and these do speed up the work if there is a lot of honey to extract. You will need a clean vessel to catch the cappings: a plastic container, with a piece of wood across the top on which to rest the frame, will be suitable. You can then support the frame by one corner and hold it vertically over the container while you slice off the wax caps. The top and bottom bars of the frame provide a good guide for the knife to run along. The odd spot in the frame where the cappings are lower can be scratched with a fork to free the honey.

Manual capping knives, in a canister of hot water. The knives are used to slice off the caps of the wax cells to let the honey out.

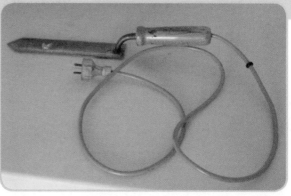

An electric capping knife

Collecting the caps' wax

A little honey will fall into the container with the wax, but this can be sorted out later. Put the uncapped frame (with both sides uncapped) into the extractor, and uncap a second frame – most small extractors hold two frames. It's not essential, but an extractor will make the job quicker while still being a low-technology affair: there are centrifugal extractors that can be worked by hand. The extracted honey will be passed through a filter afterwards to remove bits of wax.

The uncapped frames are put into the extractor, which is spun by turning the handle at the top. The centrifugal force extracts the honey from the frame, one side at a time, and throws it to the sides, where it runs down and is collected in the bottom of the extractor. (Commercial beekeepers use electric centrifuges that hold a greater number of frames.) To avoid damaging the frame, extract about half the honey from the first side, then

Cutting the caps off the combs with an electric capping knife in preparation for honey extraction.

reverse the frame to extract all the honey from the second side. Then reverse the frame again to finish the first side. Don't be too enthusiastic; spinning the frames too fast can damage the comb.

If you have no extractor, or want to use traditional methods, the honey can be extracted by hand. The traditional way to separate honey and wax is to sieve the mixture through muslin into a bucket. Make a square frame, suspend it over a large vessel, cover it with muslin … and you have your extractor. Holding it over the muslin, scrape the honeycomb with a large spoon. The wax and honey will fall onto the muslin and the honey will drip through it into the bowl, provided it is warm enough.

Honey from the centrifugal extractor may still contain bits of wax and the odd unfortunate bee, so impurities need to be removed. This can be done by settling or filtering, or both. If honey is left in a vessel to settle, the lighter wax and other impurities will float to the top and can be skimmed off, or the honey can be run out through a tap at the bottom.

Alternatively, it can be filtered through a fine mesh into another bucket before bottling.

A manual honey extractor that fits four frames. The frames whirl round inside so that the centrifugal force throws the honey to the walls of the machine where it collects at the bottom.

smaller, two-frame manual extractor. Note that you are not allowed to use an extractor outside. (Nic Moore)

Bill Ringin operating a manual extractor. If you are a small-scale beekeeper, the laundry is a great place for this job.

Drawing off the honey from the extractor

During processing the honey should not be left exposed to the air for very long because, as we have seen, it attracts moisture, which may cause it to ferment.

When the honey has been extracted, the combs are still on the frames but they are sticky – wet with honey. When they are put back in the hive they came from, the residues of honey will give the bees additional food.

They will quickly clean the frames, which can then either be left with the bees for refilling or removed for storage if the colony does not need them. Or you can start again; melt down the wax (how to process wax is explained in chapter 11) and then clean up the frame and fit it with a new foundation. (Some people like to chew the wax cappings as a sort of nutritious chewing gum. The wax is harmless to the stomach but it passes through your system undigested.)

Storing honey

Your honey harvest should now be in clean and sterile jars, either plastic or glass, or in 1-kg plastic buckets. If you intend to sell it, your labels will need to meet the regulations. The label should (usually) display your name and address, the honey variety, the net weight in the container, and basic nutrition information per 100 g – but check with the relevant authorities in your area before you have the labels printed.

There are a few facts to remember about storage, but in the right conditions honey will keep for a long time.

Most importantly, the containers must be airtight, to keep out ants, moisture and any unwanted odours the honey might absorb in the storage area. The optimum temperature range is the same as for cheese and wine, 10–16°C. Heat is detrimental to honey. The quality goes down as the temperature rises.

Alternative health practitioners and others who believe in the medicinal properties of honey often say that you should use 'raw, unpasteurised honey', which seems to imply that honey from supermarket shelves is heat-treated and therefore inferior. This may be the case if it is imported, but I have investigated two well-known brands of Australian honey on sale in our local supermarket and, according to the online information, both appear to be raw and unpasteurised. However, they have been filtered. (There are also special brands of medicinal honey, see chapter 12.)

If honey is granulated, that is, rough crystals have formed, it is hard to handle and may be heated. One source suggests that heating for up to two days at 40–50°C will not damage the honey. A (warm) water bath is the usual method.

Honey should be stored in the dark because it changes colour when exposed to light. A honey that starts off as almost white will darken considerably during storage in the light, especially at room temperature.

Honey quality

How do you know it's good honey? First of all, it should look clean, with no bits of wax or bee hairs or crystals. Nobody wants debris in their honey, but this kind of fault can be fixed by filtering, although, as we shall see, there is also a demand for unfiltered honey as it comes from the hive. It should be consistent and not set in layers.

The old-fashioned qualities of aroma, flavour and consistency are the most important in home honey production, even though scientists can now analyse the difference between honey types. The fragrance of honey should be obvious when you take the lid off the jar; a lack of aroma means that the honey has been heated and the volatile oils have evaporated. Flavour is associated with aroma and can vary considerably, depending on the nectar source. It's interesting to compare the flavour of different honeys.

Fresh honey should flow from a knife in a continuous stream and not break up into separate drops. It should form a bead when it falls. When poured it should form layers that soon disappear; if it is thin and just runs into a pool, this means that the water content is too high and the honey will not keep very long. Viscosity is a measure of the honey's keeping quality.

Honey colour can vary greatly as we have seen; it is measured by a Pfund grader, used to measure shades of yellow.

Possible contaminants of honey

It's as well to be aware of the possible sources of honey contamination, even though your backyard honey may not be in danger from any of them. Some of them have been covered elsewhere.

- If you harvest combs that are not properly sealed there may be too much moisture in the honey.
- The use of old combs can cause problems. They may contain honey from last year that could contaminate the new honey with yeasts. Dark, old combs can darken the colour of the honey and give it higher acidity.
- If the hives are in an industrial area, there may be pollution from the environment, including agricultural chemicals.
- Antibiotics and other drugs, if used to combat or suppress bee diseases, might get into the honey.
- Too much smoke used during harvesting can get into the honey.

Other products from the hive

Propolis

As if the poor little bees didn't have enough work to do, they are sometimes encouraged to make more propolis for harvesting. Propolis is made from resin gathered by bees from the buds and flowers of some plants; it acts to protect the buds. The bees scrape it off with their mandibles and take it back to the hive in pellets on their hind legs. The resulting sticky mixture of plant resin, saliva and a little wax is used as a glue to help to seal cracks in the hive. It also fights disease, since it contains a high level of flavonoids, common plant pigments which act as antioxidants. The Latin name means 'defender of the city' and this is what propolis does for the bees. They clean out the cells with it and also cover the bodies of any invaders that they have killed but can't get rid of, such as mice.

As gathered, propolis is a sticky black or brown substance with a distinctive aroma. It is collected by beekeepers on a 'propolis mat' on the upper side of which the substance is deposited, and can be scraped off with the hive tool.

The use of propolis for its medicinal properties is supported by science and, although it is not as well known as royal jelly, its use is growing.

Pollen

Bees collect pollen for a protein food, and humans collect pollen from bees, using a pollen trap, which is designed not to catch all the pollen. There are many pollen trap designs on the market. Pollen varies in colour, flavour and aroma depending on the plants it comes from. It is a high-protein food and is sold in health-food shops in various forms. Some beekeepers save it for the bees in winter, or when their diet is low in protein.

Pollens vary greatly in protein content, and bees seem able to detect this and prefer high protein pollens.

Royal jelly

You may not want to collect royal jelly, but it's interesting to have a look at it.

Royal jelly is a milky fluid, secreted by glands in the head of young worker bees. The substance has an almost mystical reputation and, judging by the number of scientific papers about its ingredients, there is still a great deal of interest in it. Royal jelly is the high-protein mixture fed by the nurse bees to emerging larvae for their first three days, and fed to the larvae in queen cells in much greater amounts. It is royal jelly that makes a larva into a queen bee, an extremely fertile insect that lives much longer than the workers. This is probably why there is a market for it, though there seems to be no real scientific evidence of its alleged positive effects on humans, in spite of widespread anecdotal 'evidence'. 'Wellbeing, euphoria and rejuvenation' are very hard to quantify! Royal jelly is defined as a dietary supplement rather than a medicine.

Royal jelly was first promoted in France in the 1950s. It seems the jelly is made up of water, protein, sugars and fatty acids with Vitamin B and trace minerals. It is not recommended for people with allergies or asthma.

What do scientists think of royal jelly? There is some scepticism about the design of studies of its effects. Dr Maleszka and colleagues at the Australian National University (2008) have demystified the way in which royal jelly can turn a worker egg into a queen bee; they believe it alters how the bee's DNA is read, but this will only work on bee DNA, not on that of humans.

Beekeepers who harvest royal jelly rob the larvae cells and have to manipulate their colonies, stimulating them to produce more queen cells from where it can be harvested. This is not really a job for a beginner and I can't imagine that many backyard beekeepers would want to embark on the process. At three days old, the queen larva is floating in royal jelly and at this point the jelly is harvested, before it can be eaten by the developing larva.

Royal jelly should be kept at or below 5°C. Often it is frozen when harvested and kept in the dark, sometimes with honey or beeswax added so that it keeps longer. Commercial producers often freeze-dry it. It is used in food supplements and in some cosmetics.

Royal jelly contains 67% water, 12.5% crude protein and 11% sugars, plus fatty acids, enzymes, antibiotics, Vitamin B and traces of Vitamin C.

CHAPTER 10

Using Honey

I eat my peas with honey;
I've done it all my life.
It does taste kind of funny
But it keeps them on the knife.

Anon

Nectar into honey

The honey-making process starts as soon as nectar goes into the bee's honey sac. Glands in the head and thorax produce enzymes which mix with the nectar as it passes from the mouth parts into the honey sac. Then the process begins whereby sucrose is split into fructose and glucose, which are sweeter. Honey is transported through the hive by passing from the foragers to other workers and this increases the enzyme content.

The honey is completed by a ripening process and by evaporation of the water until the honey is about 18% water. A colony can evaporate several litres of water during the night after a good day of foraging. The warm, moist air in the hive is fanned out by bees' wings and as they work, the moisture sometimes condenses to form a stream of water just outside the hive. You can see this on a cool morning or evening, after a busy day.

The honey is sealed into the combs when it is ripe, a few days after delivery. At this stage the moisture content is below the level at which honey will ferment or spoil. The bees know just what they are doing and we have to wait for them to do it.

The exact composition of honey varies according to the plant species on which the bee forages; on average it contains about 18% moisture. About

75% of honey is in the form of invert sugar, which can be easily assimilated by the digestive system. The sugars involved are a complex mixture and many of them are not found in nectar, but are produced during the ripening and storage stage, aided by bee enzymes and the acids found in honey, the chief of which is gluconic acid.

The other reason why honey is a more valuable food than sugar is the presence of enzymes and also of vitamins and minerals, but the content is very low. The enzymes include one that produces hydrogen peroxide, which protects the stored honey from spoilage and kills bacteria. Hydrogen peroxide is used as an antiseptic but it is less effective than honey. (More about this later.)

Cooking with honey

People are now more aware of their blood glucose levels, especially since diabetes is very common. They are looking at the GI (glycaemic index) rating of carbohydrate food, which classifies carbohydrates on a scale of 1 to 100 according to how fast they raise blood glucose levels. Low GI foods are rated 1 to 55, moderate from 55 to 69 and high GI foods have a rating of more than 70.

It used to be thought that simple carbs such as honey would raise blood sugar fast, but it now appears that the more complex foods such as bread

A nice harvest of Gippsland honey

A jar of Nic's Bambra Cross 'pure Otway honey' (Nic Moore)

have a bigger effect. And we're told that a slow trickle of glucose into your bloodstream 'keeps your energy levels balanced and means you will feel fuller for longer between meals'. You can find out more on www. glycemicindex.com. According to the University of Sydney, honey has a low to moderate GI rating. Interestingly, the rating varies with the variety of honey.

Most of us will eat honey on toast or perhaps on porridge, but there are dozens of recipes with honey as an ingredient. Honey in cakes improves the keeping quality and helps to keep them moist.

Honey as a sugar substitute

Honey can be used in many recipes instead of sugar and since it's sweeter, you should use about half the amount of honey. Some recipes recommend replacing one and a quarter cups of sugar with one cup of honey. It's hard to weigh and perhaps best measured with a spoon dipped in hot water. If you are substituting just a little of the sugar in a recipe with honey it will be more effective to use a dark, strongly flavoured honey to get the effect. If all the sugar is to be replaced, use a lighter honey. Using too much honey can result in a very dark brown cake.

Honey as we've seen is acid so, unless the recipe includes something acidic like sour cream, add a little baking soda to neutralise the acid.

A baker's tip: measure the oil or fat for the recipe first and then the honey in the same container. It will slide out more easily!

Drinks with honey

Honey drinks can be hot or cold. It can be added to tea and coffee instead of sugar.

Hot chocolate

Make a chocolate mixture by mixing in a pan half a cup each of honey, cocoa and water. Heat gently until the mixture thickens and add a little vanilla. To serve, stir up with three cups of hot milk.

Iced tea

Make tea with tea bags and boiling water. To three cups of the tea, add three cups of apple juice and stir in honey to taste, then chill.

Honey fruit syrup

This is an old recipe, rather like a cordial. It is diluted with water to make a drink.

If you have plenty of home grown fruit it would be a good way to use some of it. Press the fruit and filter the juice, then add honey, rather more honey than juice. Heat up the mixture to about 65°C and bottle. My family used to make it with elderberries for a winter cold cure.

Baking with honey

Most children have a go at making Honey Joys. Butter, sugar, margarine and honey are heated until they froth and then corn flakes are added. The mixture is spooned into patty cases and baked in a cool oven for 10 minutes.

Fruit cakes

Use your favourite fruit cake recipe and use honey instead of sugar, remembering that honey is denser than sugar and therefore sweeter for the same volume.

Honey biscuits

125 g butter
3 tablespoons honey
1 egg
half cup chopped dried apricots
2 cups plain flour
2 tablespoons cocoa

Mix butter and honey in a bowl, beat in the egg, stir in apricots, sieved flour and cocoa. Mix well and use a teaspoon to make into little balls, which can be flattened with a fork just before baking. Bake at 280°C for 10–12 minutes.

Muesli

A tasty toasted muesli can be made by mixing rolled oats, bran, chopped nuts, desiccated coconut and seeds. Mix them in a bowl and add mixed honey and vegetable oil to coat the dry ingredients. Then bake on baking trays, stirring from time to time, about 30 minutes or until toasted. Turn off the oven and add some chopped dried fruit, then leave in the oven for another 20 minutes. Stir several times while the mixture is cooling to crunchiness and when cold, store in an airtight container.

Mrs Beeton's honey cake (1st edition)

½ cup of sugar
1 cup of sour cream
2 cups flour
½ teaspoon carbonate of soda
honey to taste

Mix up the sugar and cream, add the flour and enough honey to flavour the mixture nicely. Stir well, add the carbonate of soda, beat well for five minutes, bake for about 30 minutes and 'let it be eaten warm'.

Cooking honey with meat

The traditional use of honey with meat is as a glaze. Lamb or pork can be partly baked, then a glaze of honey plus pineapple juice is poured over the roast and it is returned to the oven. Honey can also be used in chicken recipes and 'honey chicken' knows many variations.

Honey chicken

Coat chicken pieces lightly with flour and place in a casserole dish. Mix a packet of French onion soup with 3 tablespoons honey and three quarters of a cup of water. Bake in the oven at 160°C for about an hour and a half.

Roast lamb

The leg or shoulder of lamb is cooked in an oven bag, with a mixture as follows:

2 tablespoons warm honey, 2 tablespoons olive oil, 1 tablespoon lemon juice, 2 cloves of garlic, peeled and sliced.

Alcoholic drinks made with honey

Mead

Mead was probably the first alcoholic drink made, along with several other fermented drinks made with honey, such as metheglin and cyser. Some people think that mead was a milestone in the transition from a hunter -gatherer culture. Archaeology dates it to at least 7000 BC and it turns up independently in many cultures across the world. In the colder northern countries where grapes do not grow easily, it was the alcoholic fermentation of choice and is simpler to make than beer, which came later.

Because it was not easy to make a consistent, pleasant-flavoured mead, the drink tended to fall out of favour in more sophisticated times. The current resurgence of popularity is in part due to the better understanding of the fermentation organisms and the process. But mead is still not entirely predictable.

Mead has many local names. 'Bracket' is one, which became 'botchet' on the moors in Yorkshire, where a scented, strong mead was brewed from heather honey.

Mead is made by fermenting honey and water with yeast, fruit acid and tannin.

English recipes suggest that dry mead should be made with light honey and sweet mead with dark honey. This is supposed to be because of the colour of sherry; dry sherry is a pale yellow – and so people expect a sweeter drink with a darker colour. The finished product will have about the same alcohol content as grape wine, 11–12 %.

If you would like to try one of these recipes with your own honey, use standard brewing equipment from a home brew shop. But be warned, the process is slower than making beer. Mead takes at least 6 months to finish fermenting and will be better after another six months. This means that a glass container will be much better than a plastic one.

The type of yeast to use is the subject of debate and there are whole internet websites devoted to arguments about yeasts for various types of brewing. Some people say that a beer yeast is fine, others insist that you should not use beer yeast whatever you do. They say wine yeast is better, but you can buy yeasts specifically for mead making.

The more sugar is fermented by the yeast, the drier (less sweet) is the drink and you can choose a variety of yeast to give you a dry or sweet mead.

A traditional horn of mead

But the best option to start with might be to go for a medium yeast and vary the honey content to taste.

It is now standard practice to sterilise vessels and bottles with a chemical (metabisulphite) before you brew, but our ancestors didn't do it. There is the chance you might lose a batch if the wrong bugs get into the brew, of course. The alternative is good old-fashioned boiling water, followed by a quick dry near the wood stove or in the sun.

Another question: should you boil the honey? If it has been well filtered this should not be necessary. The argument for boiling is that scum rises to the surface and can be skimmed off, and if left in the mixture this might make the mead cloudy.

Boiling might spoil the honey flavour and will certainly destroy those mysterious enzymes.

If you use dried yeast it has to be reactivated before you start. Put it gently in on the top of a cupful of honey and water and leave it to start bubbling. The packet might tell you how long to leave it before use.

Basic mead ingredients

4.5 L water (demi-johns, fermenting jars are usually one gallon)
honey for dry mead, 7.5 kg
honey for medium mead, 8–9 kg
honey for sweet mead, 10–12 kg
2 lemons or limes
yeast, as described above

Mix the honey with some of the water, which may be easier if the honey is warm, but the liquid should not be hot. Tepid is the right temperature for yeast. Pour it into the jar; juice the lemons and add them.

Put the yeast starter in and fill up to the shoulder of the jar with water. You need some space for fermentation. Fit an airlock and keep the jar in a reasonably warm place.

In about six months, rack the mead. Siphon off the liquid, leaving the sediment (lees) at the bottom. When it has been really clear for a few weeks it can be bottled, but beware of bottling if there is a chance it may still be fermenting. The bottles might explode. Label the bottles with name and date and record your recipe. When you make the next batch, you may want to amend the procedure in the light of your results, or you may want to repeat the process exactly.

There are various variations on mead, which go by different names.

Sparkling mead

The sparkle can be added by re-fermenting the mead with more honey and bottling before fermentation is complete, which can be risky. Bottles can explode.

Cyser

In this recipe the water is replaced by apple juice. A mixture of sweet and sour apples is recommended. This time the lemons will not be needed, as the acid will be provided by the apples.

4.5 kg apple juice
1 kg honey
yeast, activated as for mead

Mince and press fresh apples; put the juice in the jar and leave overnight. The next day, siphon off the juice from the pulp. Heat up the honey with some of the juice (some say bring to the boil, others say not) and mix into the apple juice. When the mixture is about 25°C, stir it up and add the yeast.

Fill the gallon jar to the shoulder and add an airlock as before. Leftover mixture can be added to the jar when the vigorous fermentation is over. As with mead, cyser will improve with keeping.

Metheglin

1.5 kg honey
1 teaspoon whole cloves
1 teaspoon whole allspice

1 cinnamon stick
shavings from a nutmeg
2 tea bags
all-purpose wine yeast
yeast nutrient
4.5 L (1 gallon) water

This is mead with various spices added. Metheglyn is one of the few words that came down from Welsh into English and it also shows the association of honey with healing. 'Medd' is honey and 'meddyg' is 'doctor' in Welsh, hence 'Meddygllyn' – liquor doctor. It was used as a medicine. Cinnamon, cloves, nutmeg, rosemary and thyme are all included in one commercial variety.

Method Activate the yeast with water and honey in a warm place.

Wrap the spices in muslin and tie the bag with string. Boil the water in a large pot and drop in the spice bag and the tea bags. After a few minutes, take out the tea bags; leave the spices for about 20 minutes, by when their fragrance will be coming out. If you like a very spicy drink, you can leave the spices in longer.

Add the honey, stir well and simmer (don't boil) for about an hour, skimming off any scum from the surface.

When the honey is lukewarm, pour into a fermentation bin and add the yeast and yeast nutrient. Stir twice a day for several days and when the most violent fermentation is over, pour into a demijohn with an airlock. Keep warm until the fermentation has slowed down. In about three weeks a sediment will have developed.

To get rid of this, rack off into a clean vessel and airlock again and repeat this process when fermentation has stopped. Bottle when the wine is clear.

Hippocras – spiced wine or mead

Spiced wine was a favourite with Queen Elizabeth I and it is still made, and drunk hot in the winter in cold-climate countries. One commercial product contains ginger, cinnamon, clove and rosemary.

Sack

References to sack crop up in historical novels and it sometimes meant a very sweet sherry, but it can also be used for extra sweet mead

Honey beer

This is apparently an African speciality. It has to be consumed as it is made as it won't keep. One recipe suggests 5 litres of honey with 18 litres of water (you would need to be arranging a large party), with 6 teaspoons of dry baking yeast, kept in a warm, dark place and ready in about a week. Perhaps you should scale down the amounts for domestic consumption.

Honey liqueur

The liqueurs derive their kick from added spirits such as gin or vodka. Some of the old recipes sound drastic, but one to try might be the following: macerate dry orange peel in the alcohol of your choice. After two weeks, filter the liquid and add honey dissolved in an equal amount of water.

Hazards with honey consumption

In general, honey is a very safe food but there are hazards to remember, apart from the risk of contaminants mentioned in the previous section.

It is recommended that children under a year old should not be given honey. There is a very slight chance of botulism from spores which can be present, the dormant endospores of a bacterium, *Clostridium botulinum*. When the bacteria develop they produce toxins. Infants' digestive systems are not able to destroy the spores, which older children and adults can usually do.

Another potential problem is with honey from the nectar of certain plants, which can cause dizziness and nausea and less commonly, low blood pressure and shock. The plants include rhododendron, some laurels and some azaleas. There are stories going back into antiquity, including one of the Roman general Pompei's army in 69 BC being given this honey as a strategy by their enemies. There is speculation that an ancient Greek bee cult used 'mad honey' to induce a trance. In Britain, rhododendron is now very unpopular as an invasive exotic, smothering the native woods, and honey intoxication is listed as one of its sins. It will be as well to site your hives well away from rhododendron bushes. Tutin honey has been mentioned above, a hazard that occurs in certain parts of New Zealand under certain conditions.

CHAPTER 11

Beeswax

Night's candles are burnt out, and jocund day
Stands tiptoe on the misty mountain tops.

William Shakespeare

Until the twentieth century, you had to be rich to burn out night's candles very often. Down the centuries, beeswax has always been a luxury product. Only churches and the houses of the wealthy were lit by beeswax candles, which give a mellow light and a soft perfume. While the wealthy enjoyed them, ordinary folk used them very sparingly, if at all. Poor people who did not keep bees made do with tallow dips. They dipped rushes in mutton fat; the fat went rancid over time, the light was poor and the resulting smell was most unpleasant.

Pure beeswax candles are still quite expensive, but they are a 'green' choice because unlike paraffin wax candles, they are carbon neutral. The quantity of carbon dioxide released while the wax is burning has already been removed from the environment by the bees when they produced the wax.

Even the light emitted by a beeswax candle is superior, say the enthusiasts. It's a warm glow in the same colour spectrum as the sun.

Beeswax is a renewable resource, a totally natural product with its own colour and fragrance. Australian beeswax is supposed to be (and probably is) the purest you can get. Candles sold are often not of pure beeswax; they have to be only 51% beeswax to be labelled as such. But if you produce your own, from your own bees, you will know that they are the real thing.

Candlelight is often used for special occasions, for its mysterious ambience. But do we know what we are doing to the air we breathe? Paraffin candles often contain artificial scents that create black soot. It's said that pure beeswax candles purify the air, producing negative ions, which clear the air of allergens, smoke and odours. The ions attach to noxious particles and, being heavier, drag them to the ground. Beeswax is said to be the only fuel with this property, which is apparently found in free-flowing water such as waterfalls and oceans. Beeswax candles burn more slowly than those made with paraffin wax, and they don't smoke if the wick is the right size.

The colour of wax varies from a lovely dark yellow to an almost white colour when it has been bleached by the sun. Although wax from the brood box looks dirty and unusable, this wax provides the best deep yellow colour to make candles. Wax from the super is lighter and not as attractive as brood box wax. Wax left in the dark gets a coating of grey bloom that provides an aged look to the candle – or to foundation. The bloom can easily be removed by leaving the candle in the sun for a few minutes, wiping with water, or gently blowing a hair dryer over it. If left on foundation, the bees will quickly remove it and it won't harm them.

We should value this product. It takes the bees several kilos of honey to produce one kilo of wax. As we saw earlier, young worker bees have eight wax glands on the thorax. When they are older they go out foraging and the glands gradually disappear. From 12–18 days old, they hang in a group and produce clear wax scales that are then worked in the mouth-parts and become softer and opaque. New wax is nearly white, but it is often coloured by pollen and propolis. Wax-producing bees cluster like a hanging chain and this gives them a line on which to build.

Beeswax has many other uses apart from candles. The Romans used wax for writing tablets. They scratched letters into the surface with a stylus and the wax could later be smoothed over and the tablet used again. They also used beeswax in buildings to make walls waterproof, which indicates that they must have gone in for beekeeping in a big way.

Wax, being plastic, can be warmed and used to make moulds for jewellery, and for modelling, although in some cases it has been replaced by modern plastics. The ancient technique is still used. A wax model is filled with plaster; when the plaster has set, the wax is melted and the plaster mould can then be used for molten metal. It is also used for dental moulds.

Beeswax polish is used for furniture to give it a mellow glow. It can be mixed with various other substances such as linseed, tung oil and turpentine to be used as a furniture finish.

Traditional cheese makers use wax to coat some finished cheeses, for example large Cheddars. The wax may be coloured and is the perfect protection for cheese that is allowed to mature over months or years. Beeswax is allowed to be used as a glazing agent for food.

Several types of instrument makers use beeswax. Accordion makers blend it with adhesive and didgeridoo players use it as a mouthpiece. It is a sealant and a lubricant in many trades and used for threads and bowstrings.

Perhaps the main use for beeswax these days is in cosmetics and pharmaceuticals. In 2003, a German study found that beeswax was better as a barrier cream than products made of mineral oils (see recipes at the end of this chapter). The skin does not absorb beeswax, so it forms a barrier which prevents moisture loss. If you like to make your own ointments, it is a useful base material to which can be added vegetable oils or herbs such as lavender.

Recovering the wax

As we have seen, beeswax is a valuable commodity that you can use in a variety of ways, or you can sell it to bee suppliers to make wax foundation for beekeepers.

Once the honey is extracted, you are left with sticky wax cappings. Drain them of as much honey as you can. Some people have a cappings bag for the extractor, a mesh bag the same size as a frame. They fill it with cappings and give it a whirl in the centrifuge to remove the honey residues.

Old combs are also a source of fresh wax if, as suggested above, you melt them down and start again. When they have been in use for a long time, the skins shed by a succession of emerging pupae reduce the size of the cell. Combs used just for honey production eventually get dirty.

The third source of wax is 'slum', the sticky stuff you scrape off the tops of frames and edges of boxes. Slum or slumgum is a mixture of propolis, wax and dirt, and provides the glue to fill any cracks in the bees' home.

Wax is separated from the other substances by heat and there are several ways in which to do it. The melting point of wax is about 64°C, but if it is heated to 85°C it becomes discoloured. Its specific gravity is less than one, so it floats on water.

Solar extractor

You can buy or make a solar wax melter: a box painted dark on the inside, with a glass lid set on a slant. The sun melts the wax and it runs into a collecting tray at the bottom, where it solidifies into a block. The residue stays in the melter. Obviously the solar melter will only work on warm sunny days. If you do use one, take the wax out after it has cooled down and solidified the first time. You could probably make a solar extractor yourself if you are handy with tools. Bob's Beekeeping Supplies website gives detailed plans of how to build a solar extractor (and the site includes much other useful information).

Heating over hot water

Take safety precautions when doing this, of course. The danger lies in the fact that beeswax doesn't boil, it just ignites when it gets hot enough and

Melt the collected wax in a large pan of hot water. *Note that the pan shown here is too full.* The melted wax rises to the top and can be skimmed off, leaving any impurities in the honey water, which can then be filtered and brewed. (Nic Moore)

can cause a serious fire. Beeswax should never be heated over 100°C, and even if it doesn't catch fire it will be damaged by overheating.

Fill a large pan to no more than two-thirds full – for safety again. Use one-third water and add one-third wax. The pan is then heated until the wax melts, which occurs at 64°C. When the pan is taken off the heat, wrap it up so that it cools down slowly. In this case the wax floats to the top and when cold, can be lifted off in a block. There will be some dirt at the bottom, which can be scraped off.

To make the operation even safer it is suggested that you should use the double boiler method and heat the wax inside a container in a pan of water. A plastic container will melt, so use stainless steel or glass.

Wax press

This is often used to extract the dark wax from old combs. It consists of a plunger with a screw inside a cylinder, in which water is heated by an electric element. The old combs and residue from solar extractors can be recovered. The wax is put into hessian bags and soaked in the cylinder for two hours. Then the water is heated gently until hot enough to melt the wax, the press is screwed down and the wax rises to the surface of the water. The water level is raised and the wax flows off into a mould.

Hot box melter

This is a stainless steel tank with an outlet at the bottom for honey. The tank is fitted into an insulated box, which is heated with a fan heater and equipped with a thermostat. The impure honey you have obtained from spinning the cappings is put into the tank and as the mixture melts the honey runs out at the bottom. Then hot water is added and the wax can be scooped off the top of the water when it has cooled and set.

When water comes into contact with wax during any recovery process, it is best that rainwater is used.

Moulds

For home use you can make do with an old ice-cream container or a cake tin, but wax produced for sale will need to be formed in a standard-sized mould. As an example, the standard mould preferred by Capilano Honey Ltd is a plastic No 4 crate that holds about 8 kg of wax. If you intend to sell wax, it will be an advantage to use a standard mould and to produce good-

A mould of pure wax (Nic Moore)

quality wax. Pour into the moulds just before the wax sets, to prevent cracking, and cover the moulds to allow the wax to cool slowly. If you decide to compete with your wax at an agricultural show, it's important to get a smooth finish. To do this, put the mould into hot water, cover and allow to cool slowly without any vibration.

Smaller moulds can be used too, because small quantities of wax may be needed. You can use bits of beeswax to help you to thread a needle, to lubricate zips or drawer runners, and to wax fishing lines. You can also use it to get a spiky hair-style effect or to give a shine to your hair.

Clean wax will be your aim, whether you intend to sell the wax or use it yourself, although you can make good candles from dirty wax. When wax is poured after melting, the first and the last portions will contain small bits of debris. These are usually kept and put through with the next batch. Iron, brass, copper or zinc containers may discolour the wax.

If you decide not to extract wax from old combs or cappings, remember disease control and make sure that bees have no access to the material. Burning is the best method of disposal as it will kill any spores present.

Candle making

The soft perfume of a burning wax candle, with a hint of pollen and honey, complements its mellow light. You can also add other perfumes such as vanilla or lavender oil, dyes for colour and even flowers for an interesting effect.

If you want to make the beeswax go further it can be mixed with paraffin wax. Even with no more than 25% beeswax, the candle will retain its natural fragrance. This and all the equipment needed for candle making can be obtained from candle-making suppliers; it is a popular hobby.

There are three ways of making candles: dipping (as used to be done with mutton fat), moulding, and casting. In all cases, the diameter of the candle should be appropriate to the gauge of the wick used – the fatter the candle, the thicker the wick should be. If a wick is too small for a candle, it will burn a deep cavity and if it's too large, it will smoke while burning. If you enjoy experimenting, you can make a test bed and measure the performance of different-sized wicks.

First, the wicks are cut to length and waxed. It is wise to buy proper candlewick rather than using a piece of old string. The lengths are submerged in very hot wax until it stops bubbling, which it does when all the moisture has been driven out of the wick.

Candle dipping

Heat the wax gently in a double boiler until it melts. This is when you add colour and scent if required. Cut the wick to the right length for the candles, plus a little more. Dip the wick in the wax for a few seconds and then take it out (you can try two at a time). Let it cool for a minute or so between dippings and repeat until you get to the right thickness. Some 15 to 20 dips should ensure that a satisfactory thickness of wax sets around the wicks.

Candle moulds

Moulding of course involves pouring the melted wax into a mould and allowing it to cool. Moulds can be made of metal, with a hole in the base for the wick so you can keep it taut. They may have a seam which has to be shaved off the candle at the end.

Plastic moulds are usually in two pieces. Rubber moulds are not suitable for beeswax. You can use waxed milk cartons or recycled plastic containers, which can be torn away when the candles have set.

Glass moulds are easy to use and if you like fancy patterns on the candles, this is the way to go. The mould is rinsed with a detergent solution and then drained, leaving a film of detergent, which will allow the mould to release the candle when it cools. The wick is threaded into the mould and the hot wax is poured in carefully down the wick, keeping the mould upright – it should be firmly propped before you start. Put the wax back in the water bath for later topping up as soon as the wax starts to set around the edge of the mould.

Leave the candle in the mould until the next day and put it in cold water, when the candle should float out.

Candle casting

This is a way of making tapered candles. The wax is poured down a wick, twisting the wick while pouring, a tricky operation. These days, casting is used to mean the use of a temporary mould such as metal foil, which is not used again.

The candles you make can be polished gently with a damp towel and finished off with a soft cloth – beware of scratching. They may develop a light bloom over time, but this can soon be rubbed off.

Candles are usually stored at room temperature, although some people think they are better off in a freezer. In a hot summer your candles may soften and bend if not kept cool. Tapered candles should be stored flat.

Some beeswax recipes

Making hand creams and other cosmetics is another popular hobby, and with a supply of your own beeswax you have a head start. People with very sensitive skin often like to know what they are using.

Moisturising cream (from rachelssupply.com)

125 g sweet almond oil
30 g beeswax
60 g water
10 drops Vitamin E oil
10 drops lavender oil

Heat the oils in a double boiler with the wax, mix and then stir in the water. Add the Vitamin E oil and keep stirring until the mixture cools down. Pour it into jars before it solidifies. Lavender oil is healing and disinfecting, as well as giving the cream a pleasant fragrance.

Solid perfume

3 parts sweet almond or jojoba oil
2 parts white beeswax
1 part fragrance oil

Melt all the ingredients over a double boiler and mix well. Pour into containers and leave to solidify.

Beeswax soap

There are many recipes for a gentle soap made with beeswax, or beeswax and honey. Making soap is yet another craft hobby and no doubt our ancestors, who had to make their own soap, would smile at the amount of perfume to be used. The ingredients are beeswax, sodium hydroxide (lye), olive oil, tallow, coconut oil, water and a dash of scented oil.

Beeswax with wood

Wooden drawers run more smoothly when coated with beeswax and the trick is to rub a piece of wax along the runners. It also works on old sliding windows and sashes.

It is a natural substance to use on bare wood and the traditional beeswax, linseed oil and turpentine mixture preserves wooden beams and makes a good wood finish where the wood will not get heavy wear. Use equal quantities of the ingredients; heat up the wax until it melts, then take away from the heat and mix in the linseed oil and turpentine.

This mixture works better if it is slightly warm when used and applied with a soft cloth and left to dry. The surface is then buffed with a wool cloth.

Nic's home-made furniture polish, made with natural turps and wax.

Beeswax and metal

Melted wax will help you to free a rusted nut and make screws drive more smoothly.

Beeswax and turpentine will help to stop the oxidation of bronze which is caused by damp air. Buff the metal to create a hard coat of the mixture.

Rub copper with softened beeswax to give it a patina.

Beeswax and leather

Leather can be waterproofed with a traditional mix of beeswax, tallow, and Neatsfoot oil (usually obtainable from horse equipment suppliers). Combine equal parts of these ingredients and apply warm with a cloth to coats or boots.

Beeswax and stone

Counter tops of stone or polished concrete can be given a lustre by rubbing them with melted beeswax.

CHAPTER 12

Apitherapy, the Medicinal Use of Bee Products

Weave a circle round him thrice,
And close your eyes with holy dread,
For he on honeydew hath fed,
And drunk the milk of Paradise.

Samuel Taylor Coleridge

From the mists of time and the magical origins of medicine, the tradition lingers that honey is a cure for nearly every ill. There is such a long catalogue of benefits that it makes one cautious; can it all be true? Probably not, but the proven uses make an impressive list. There has after all been a medical connection for a very long time. For thousands of years, advisers of all persuasions have recommended honey for health.

You may wonder why, if it is such good medicine, honey is not more universally used as such. During the last 50 years we have become addicted to antibiotics, which have proved to be a mixed blessing, along with other strong pharmaceuticals. Until recently, many natural remedies had been abandoned by all but the diehard herbalists in favour of faster cures.

The answer may be that honey's folksy image and its long history have branded it as alternative and therefore suspect. It is also very cheap compared to modern drugs, it cannot be patented or licensed universally and it can't be copied! So big companies have not promoted honey as a medicine. But there are many well authenticated stories of doctors who, once they have tried honey, have used it extensively to treat their patients.

In fact things are changing: the medical use of honey is on the increase and there are now marketed brands of medical honey. When your bees start producing honey for you, your friends may recite the medical benefits because there has been a recent revival of interest in more natural therapies. This is part of a worldwide swing away from strong chemicals and a search for less aggressive treatments.

In some ways it's not surprising that the tide of public opinion should now favour natural products. Old-fashioned folkways could never compete with big corporations until people began to feel that the modern wonder drugs were not all they seemed and that in the general wellbeing of populations something was missing. So a few people turned to honey and milk instead of pills for a good night's sleep.

Honey is free from additives and contains only inbuilt, natural preservatives. It is relatively free from any adulteration; bees that encounter pesticides will usually die before they can get back to the hive and pass on the poison in the honey (but see the section on contamination). The good thing is that it won't hurt you, so there is no harm in trying the honey cure.

Many sources suggest that 'raw honey' straight from the hive should be used medically and as a beekeeper, you will have access to your own supply. They say that the honey should not even be strained. Unstrained honey can be cloudy and thus less attractive to the consumer, but it will have retained all its benefits. The bits of wax and propolis, and the odd grains of pollen may all have a part to play in promoting health, or so say the enthusiasts. Honey is heavier than its impurities, so if liquid honey is allowed to settle in a tank for a few days, much of the wax, pollen and so on will rise to the top and can be removed.

There are folklore and historical references to the use of honey as medicine, going back thousands of years (and I don't suppose the ancients would have strained their honey). As often happens with traditional wisdom, some of the claims are now being confirmed and supported by clinical tests. Naturally enough, the folklore claims extend much more widely than the scientific evidence – so far!

I thought it would be useful to review a selection of both, but a word of warning: if you decide to try honey as a medicine, take proper advice. This section is intended to provide background information, not to prescribe for any disorders. But honey is not a dangerous substance, so unless you need urgent treatment it should be safe to try it for common ills.

Antimicrobial action of honey

It is now accepted by scientists that honey can kill micro-organisms, which gives it many uses in medicine. Jeffrey and Echazaretta (1996) in a paper about the medical uses of honey from the University of Yucatan, Mexico, have reviewed several reports.

Three thousand years ago, the Arab physician Avicenna recommended honey for the treatment of wounds. There may be several reasons why this works. The 1996 paper tells us that one of the enzymes in honey, glucose oxidase, manufactured by the bees, breaks down glucose into gluconic acid and hydrogen peroxide, a chemical often used to kill bacteria. In the hive, hydrogen peroxide stabilises the ripening honey and improves its keeping quality.

It seems that a solution of hydrogen peroxide as an antiseptic is less effective than the naturally occurring, slow release preparation, honey. Hydrogen peroxide on its own is effervescent and breaks down quickly into water and oxygen. But when honey is used as a wound dressing, hydrogen peroxide is produced slowly and acts as an antiseptic.

Because honey is hygroscopic – absorbs moisture – it is thought that bacteria exposed to honey lose moisture from their cells to the honey, thus disrupting their balance and drying them out so that they die.

There are some other factors in honey that inhibit the growth of well-known pathogens such as *Escherichia coli* (E. coli), *Salmonella* and *Staphylococcus aureus* (golden staph). Most of the organisms that can survive in honey are sugar-tolerant yeasts, which is just as well for mead makers, who depend on the yeasts to convert the sugar in honey into alcohol.

Organic acids may play a part and some of the antiseptics come from the flowers, the source of the honey. In clinical trials, honey from Mimosa and Eucalyptus species produced by Africanised honey bees had the greatest antibacterial activity. New Zealanders say the same about their Manuka honey, which is described as 'hospital grade' honey. Australia has a high antimicrobial activity honey sold by chemists, called Medihoney. It is harvested from a coastal tea tree (*Leptospermum*), locally called Jelly Bush. Herbs like lavender and rosemary have disinfecting and healing properties that may be passed on in honey.

The paper by Jeffrey and Echazaretta includes a variety of applications for honey.

Gastroenteritis

Infectious organisms have been inhibited by a 20% solution of honey. Sixty-six per cent of patients treated with honey were cured and 17% were improved. For infants, honey was effective instead of glucose in a rehydrating solution (but there is a warning elsewhere that honey should not be given to babies under one year because of the danger of botulism spores).

Gastric ulcers

Honey reduces the production of gastric acid and there is a report of an 80% recovery rate among 600 gastric ulcer patients treated with honey.

Burns, wounds and abrasions

Externally, honey has a long history of healing wounds and reducing infections. In the memoirs of a nineteenth century doctor, he records that he was unable to cure a patient with boils on the backs of his hands. The patient treated them himself with honey and made a swift recovery.

The texture of honey itself makes a good dressing. As you will know, a wet dressing on a wound delays healing by keeping the skin moist. Dry dressings stick to the wound and cause pain and injury to the skin when they are changed. Oily dressings can spread the infection. But honey is non-toxic, non-irritating, sterile and, above all, comfortable, say the medics. (However, I was told by one patient that she found the honey dressing on a wound to be very painful.) The treatment of wounds with honey has made them sterile and started the healing process within 7–10 days. It can reduce the healing time for burns. Honey is also very good at encouraging the acceptance of grafts. It is suggested that honey for medical use should be protected from light.

Another advantage of honey may be that pathogens don't seem to be able to develop resistance to it as they do to antibiotics (*Apitherapy News*, June 2009). In a trial, eight antibiotic-resistant pathogens were killed by honey.

Skin ulcers

Honey has been used very effectively on ulcers that are difficult to heal. In addition to the antiseptic effect, honey may also stimulate leucocytes that the body uses to fight infection.

Diabetes

Blood sugars rise more slowly after honey is eaten and one study suggested honey as a substitute for sugar for Type II diabetic patients.

Sore throats

Honey is a popular international remedy for soothing sore throats, often taken with lemon. Now we know that it could also be acting on the infection that causes the sore throat.

Mouth and gum infections

Sugar is bad for the teeth, but honey can fight infection and is beginning to be used in dentistry.

Colds, coughs and congestion

Mix equal parts of honey and ginger juice, an old Indian remedy. Honey has a soothing effect on the inflamed respiratory tract.

Digestion

Honey and cider vinegar is diluted with water to taste.

Honey for allergies

Honey contains pollen, which is the source of some allergies and it is suggested that a teaspoon of honey a day will help to build up immunity.

Respiratory problems

According to one report, the majority of people treated with honey for asthma or bronchitis had no symptoms after treatment, while up to 30% reported an improvement.

Anti-cancer properties

There is not a lot of evidence, but honey was found to have 'moderate anti-tumour effects' in rats and mice. Acids in propolis, which will be present in small amounts in unfiltered honey, inhibited colon cancers in animals.

As a sedative, to calm down the nervous system

A mixture of 25 % honey in water is supposed to be a stabiliser. Hyperactive children may benefit from eating honey in place of white sugar. It is a treatment for bed wetting in children and it helps water retention as well as calming them. Honey in warm milk at bed time is an old cure for sleeplessness and it will also soothe a cough which could keep you awake.

Migraine

Because migraine is stress-related, honey may help; dissolve a dessertspoon of honey in half a glass of warm water and sip it at the start of an attack.

Laxative

For a mild laxative, dissolve a spoon of honey in the juice of half a lemon and drink in water, first thing in the morning.

Muscle cramps

Cramps in the legs, often occurring at night, can be relieved by honey and apple cider vinegar, which supplies potassium. Two teaspoons of cider vinegar and one of honey.

Arthritis

Hot water with honey and cinnamon is recommended. A cream made with Manuka honey is sold to reduce inflammation. Although some people think that honey is useless as an arthritis treatment, experience of what works varies a great deal according to the individual, so honey is worth a try.

For the eyes

Drink honey and carrot juice to improve poor sight and watery eyes.

For eye infections, bathe the eye with honey and water. (And one I am wary of trying: honey and onion juice, 2 g each, mixed up and kept in a sterile bottle, is applied to the eyes with a glass rod and is supposed to be a remedy for immature cataracts.)

Honey and ageing

The folklore says that regular eaters of honey are long-lived and that beekeepers suffer less from cancer and arthritis than any other occupational group. But this may be due to the effect of bee stings (see the section on bee venom).

Bone mass

It is suggested that a teaspoon of honey a day helps us to utilise calcium and prevents osteoporosis.

Anaemia

Honey is supposed to help anaemia because of its minerals. Eat a banana a day with a teaspoon of honey. Or drink apple juice and honey.

Heart disease

Honey is supposed to tone up the heart and improve circulation, one tablespoon daily after food being the recommended dose. It is also supposed to alleviate cardiac pain and palpitation of the heart. However, the replacement of white sugar, sucrose, by honey may be the key here. A high sugar diet increases the risk of heart-related conditions like high blood pressure and obesity and it reduces the effects of 'good' cholesterol. It's also possible that honey can help to reduce cholesterol and CRP levels. (CRP is a blood protein associated with heart disease.)

Honey in cosmetics

The tradition of using honey as a beautifier is a very old one. Cleopatra is said to have added honey to her bath of asses' milk. Today honey is used in many skin preparations. The antiseptic effect is useful on skin and, because honey attracts moisture, it acts as a moisturiser. It's said that the consumption of honey will also improve the skin. A treatment for acne is to mix honey with a moisturiser as a base – honey is too sticky to use on its own.

If you would like to make your own honey cosmetics, the face mask is an easy one to start with and it's supposed to nourish the skin and delay wrinkle formation. Leave the mask on your face for about 30 minutes and then remove it with warm water.

Face mask

1 teaspoon honey
1 tablespoon milk OR 1 teaspoon glycerine
white of 1 egg

Hand cream

For hand care you can make a paste and massage your hands with it.

10 g honey
6 g wheat flour
4 g water

Honey can also be added to bath water.

Hair conditioner

Equal quantities of honey and olive oil are applied to the hair. The head is covered with a warm towel for 30 minutes, then the hair is washed.

Uses of propolis

As we have seen, propolis is the bee glue, made of resin, saliva and wax. It contains flavonoids at a higher concentration than those in honey and these give it therapeutic properties. Flavonoids occur in plants, are found in many fruits and vegetables and have many functions. They have been found to modify the body's reaction to viruses and even carcinogens. They are anti-allergic and anti-inflammatory.

Recent research seems to indicate that the body's response to flavonoids is to get rid of them and in doing so, it gets rid of carcinogens, so they may help to prevent cancer.

Propolis is exported to pharmaceutical companies and is quite a valuable component of participating beekeepers' income. They scrape it with a hive tool into a plastic bucket, freeze it immediately and it is airfreighted, some of it to Japan.

Uses of propolis include:
* A mouthwash for gum problems, and a toothpaste
* An ointment for cuts and scratches
* A throat lozenge.

Royal jelly

You will remember that harvesters of royal jelly stimulate their bees to produce it by removing the queen.

In spite of the sceptics, there is a variety of royal jelly products on the market, sold in health food shops and health food sites on line. It is sometimes sold in honey, which preserves it.

Proponents of royal jelly say that it's good for the very young and the very old.

It's interesting to look at the literature for and against the benefits of this substance, which probably derived its reputation from the fact that it makes queen bees.

Royal jelly comes with a warning: it may cause an allergic reaction.

Bee venom (apitoxin)

Yes, there is even a market for the substance in bee stings. Bee venom contains enzymes, proteins and amino acids and a compound called mellitin. It is not a coagulant like snake venom, but an anticoagulant. It contains some antibiotics and sulphur, which encourages the body to release cortisone. Traditionally, the venom was administered by a live bee stinging the patient, and this is thought to be the most effective treatment. The sting is more effective in summer when the bees have a high protein diet. Therapists check for allergy before the treatment by injecting a minute amount of venom under the skin. This is generally seen as an 'alternative' therapy.

Dry bee venom is now produced for medical use. Extracting bee venom is a delicate job because it can be easily contaminated and is sensitive to light. The work is done by using an electronic extractor, a glass frame carrying an electric current, placed at the entrance to the hive or in some cases, in the hive. Bees sting the glass and the operation is repeated two or three times. The operators say it's a less aggressive operation than extracting honey. It seems that the bees don't die in this case. They release the venom without releasing the sting mechanism because the barbs can't penetrate the glass. From one hive about 150 mg of dried venom may be produced per session, as bees only deliver a small amount in each sting (although this is hard to believe when you are stung).

The moisture in the venom is evaporated by placing the glasses in a dark, dry room and the dry powder is scraped off and bottled. It is then frozen for storage.

Mixed with oils such as eucalyptus, bee venom is made into an ointment for the relief of pain associated with rheumatism and muscle pains. It is also used in immunotherapy, desensitising people who are allergic to bee stings. It is a treatment for multiple sclerosis.

Keep any eye out for news items on the value of honey and bee products; no doubt more research is on the way.

FURTHER INFORMATION

The information provided here was correct at the time of writing.

Some publications

AgNote DAI/178 from NSW Agriculture

Bee Agskills, a practical guide to farm skills, NSW Department of Primary Industries, 2007

Bees and Honey, 4th ed, NSW Department of Agriculture 1964, Sydney

Benjamin, A, and McCallum, B, *A World Without Bees*, Guardian Books 2008, UK

Crane, Dr Eva, *Bees and Beekeeping – science, practice and world resources* (1990)

Crane, Dr Eva, *The World History of Beekeeping and Honey Hunting*, Duckworth, London (1999)

Dollin, Dr Anne, and Russell and Janine Zabel, *Boxing and Splitting Hives*, Native Bees of Australia Series No 9 (available from www.zabel.com.au)

Hooper, Ted, *Guide to Bees and Honey*, Blandford Press 1976 (Rodale Press 1977)

Klumpp, John, *Australian Stingless Bees, a Guide to Sugarbag Beekeeping*, Earthling Enterprises

Peacock, Paul, *Keeping Bees, a complete, practical guide*, Octopus Publishing Group, 2008

Tautz, Jürgen, *The Buzz about Bees*, biology of a superorganism, Springer 2008

Training

Here is a selection of beekeeping courses. To apply, you may have to download a course's application form from the relevant website.

Some adult education providers run beekeeping courses and if they don't, you can suggest they should. With ten potential students they might organise a course.

New South Wales: Open Training and Education Network (OTEN)
A distance/online course 'for interested people and part-time beekeepers' gives an overview of the industry and legislation. According to the course information it includes practical sessions and costs $495.
www.oten.edu.au

New South Wales: Western Sydney Institute of TAFE
A 1-day introductory weekend course 'for people interested in learning the basics of beekeeping' costs $65. Apply online.
www.tafensw.edu.au.

New Zealand: Telford Rural Polytechnic (Owaka Highway, PO Box 6, Balclutha, South Otago)
A variety of courses, some at advanced level.
www.telford.ac.nz

South Australia: WEA Adult Education (Box 7055, Hutt Street Post Office, Adelaide 5000)
Beekeeping for Beginners (a step-by-step approach to how to start the first hive): $75.
Course of 2 weekends and 3-hour field trip
www.wea-sa.com.au

Queensland: University of Queensland (Gatton College, Lawes, Qld 4343)
A weekend beekeeping course has been run every year (but not every weekend!) for the last 50 years.
www.uq.edu.au

Tasmania: Adult Education Northern Suburbs Centre (GPO Box 874, Hobart, Tas. 7001)
Runs part-time courses over 9 weeks, costing $123.20 plus up to $200 for field trips. Personal protective equipment needed. Covers what you need to know to set up your own hive.
www.adulteducation.tas.gov.au

Victoria: Bendigo Regional Institute of TAFE (BRIT Conservation and Land Management, PO Box 170, Bendigo, Vic. 3552)
Seven weeks part-time study, this course is for both hobbyist and commercial beekeepers. It covers: 'Practical and theoretical knowledge required for the establishment, operation and maintenance of a beehive and how to responsibly manage European honey bees for honey production.'
www.britafe.vic.edu.au

Government departments

Each state and territory's department of primary industries (the actual name varies between jurisdictions) provides information and advice on most aspects of beekeeping, including bee diseases, legislation, contacts, and equipment suppliers. This is also where you register your hives. Simply try a search for 'beekeeping' on each site's home page.

New South Wales: www.dpi.nsw.gov.au
This site has an excellent section on the management of honey bees.

New Zealand: www.mat.govt.nz

Northern Territory: www.dpi.nt.gov.au

Queensland: www.dpi.qld.gov.au

South Australia: www.pir.sa.gov.au

Tasmania: www.dpif.gov.tas.au

Victoria: www.dpi.gov.vic.au

Western Australia: www.agrc.wa.gov.au (search for honey bees)

Clubs and Associations

Most of the following clubs' websites contain links to other sites and up-to-date lists of clubs and associations, so if you cannot find a club close to where you live listed below, you may well find one by searching the websites listed here.

The following site has an up-to-date list of clubs in Australia:
www.honeybee.com.au

The following site gives contact details for clubs in New Zealand:
www.nba.org.nz

ACT

Beekeepers Association of the ACT
Meetings 2nd Thursday of the month at the Canberra Institute of Technology,
Heysen Street, Weston, ACT 2611
Website is a link from www.bindaree.com.au

New South Wales

Amateur Beekeeper's Association of NSW (Inc.)
The Association provides a bi-monthly newsletter, field days to improve beekeeping skills, optional insurance, and a 'swarm system' to co-ordinate members' efforts in collecting bee swarms. Members can borrow equipment and books from ABA and may be able to buy queens and nucleus colonies from other members.
www.beekeepers.asn.au

New Zealand

National Beekeepers Association of New Zealand
Jessica Williams, Executive Secretary
PO Box 10792, Wellington 6143, NZ
M + 64 4471 6254
E secretary@nba.org.nz
Their site gives contact details for branches and clubs in New Zealand, information on registering hives, and much more:
www.nba.org.nz

Queensland

Queensland Beekeepers Association
www.qbabees.org.au

Gold Coast Amateur Beekeepers Society
This website has links to a number of other Queensland clubs, including the Brisbane Amateur Beekeepers Society Inc.
http://web.aanet.com.au/%7EBees/gcabs/

Far-north Queensland Beekeepers Association
C/- Marg Mikits, PO Box 272, Mareeba, Qld 4880

Tasmania

Tasmanian Beekeepers Association
www.tasmanianbeekeepers.org.au

South Australia

South Australian Apiarists Association
www.saaa.org.au

Amateur Beekeepers Society of SA (Inc)
C/- Roy Frisby-Smith
PO Box 283, Eastwood, SA 5063

Victoria

The Beekeepers Club Inc.
Monthly meetings and beginner's courses, field trips, an extensive library, and a website with a lot of information and some very good links, as well as discussion forums at www.beekeepers.org.au/discussion_forums.html. www.beekeepers.org.au

Victorian Apiarists Association Inc.
Gippsland Apiarists Association Inc.
Secretary, Bill Ringin
PO Box 201, Moe, Vic. 3825
T 03 56 331 326
www.vicbeekeepers.com.au

Western Australia

WA Apiarists Society Inc.
WA Beekeepers Association
Up-to-date contact details for these organisations can be found at:
www.beekeepingwestaus.asn.au

Equipment and Clothing

Bindaree Bee Supplies
A family business, open three days a week, so call them before visiting.
10 Vine Close, Murrumbateman, NSW 2582
T 02 6226 8866
www.bindaree.com.au

Bob's Beekeeping Supplies
Equipment, clothing and advice (e.g. recipe for pollen feed supplement for bees), they supply kits for DIY assembly (a complete hive is $154). Call them before visiting.
79 Zigzag Road, Eltham, Vic. 3095
T 03 9439 5410
www.bobsbeekeeping.com.au

Caracell Beekeepers Supplies
24 Andromeda Crescent, East Tamaki, New Zealand
T + 64 9 274 7236
E beesupplies@xtra.co.nz

CB Palmer & Co
This internet business allows visits to their depot at Ironbark by appointment.
PO Box 298, Ipswich, Qld 4305
T 07 3201 8118
www.honeybee.com.au

Ecroyd Beekeeping Supplies
6A Sheffield Crescent, Burnside, Christchurch 8053, New Zealand
T + 64 3 358 7498
www.ecroyd.com

John L Guilfoyle Pty Ltd.
Head office: 38 Begonia Street, Inala, Qld 4077
T 07 3279 9750
(Also in Werrington, NSW, and Adelaide, SA)
www.johnlguilfoyle.com.au

Penders Beekeeping Supplies
28 Munibung Road, Cardiff, NSW 2285
T 02 4956 6166
www.penders.net.au

Stingless Bees
Russell & Janine Zabel keep native Australian stingless bees. They sell hives of Trigona carbonaria and Austroplebeia australis and have display hives of two other native species.
3597 Warrego Highway, Hatton Vale, Qld 4341
M 0404 892139
www.zabel.com.au

Other websites

Native bee research:
www.aussiebee.com.au

www.beesource.com is a community-based American website with very active beekeeping forums where you can discuss beekeeping matters with others on: www.beesource.com/forums/index.php?s=898d7994df2161761 c8b6c9d33475663

Capilano Honey, one of the largest honey suppliers, has an informative website that includes recipes:
www.capilano.com.au

The NSW government primary industries' department's website has a terrific section on the management of honey bees:
www.dpi.nsw.gov.au/agriculture/livestock/honey-bees/management

Randy Oliver, a Californian beekeeper, maintains a website that provides a digest of recent scientific information:
www.scientificbeekeeping.com

Researcher Tim Heard's work has been responsible for much of the growing interest in native bees. He has developed a small, box-like native beehive, in two halves, so that strong colonies can be split in two.
www.sugarbag.net

INDEX